SOLAR RADIATION

SOLAR RADIATION

Edited by

N. ROBINSON

Solar Physics Laboratory
Technion - Israel Institute of Technology
Haifa, Israel

ELSEVIER PUBLISHING COMPANY
AMSTERDAM/LONDON/NEW YORK/1966

ELSEVIER PUBLISHING COMPANY
335 JAN VAN GALENSTRAAT, P.O. BOX 211, AMSTERDAM

AMERICAN ELSEVIER PUBLISHING COMPANY, INC.
52 VANDERBILT AVENUE, NEW YORK, N.Y. 10017

ELSEVIER PUBLISHING COMPANY LIMITED
RIPPLESIDE COMMERCIAL ESTATE, BARKING, ESSEX

LIBRARY OF CONGRESS CATALOG CARD NUMBER 66-11405

WITH 219 ILLUSTRATIONS, 77 TABLES AND 2 SEPARATE DIAGRAMS

PRINTED IN THE NETHERLANDS

PUBLISHER'S NOTE

Owing to the untimely death of the Editor, Professor Nathan Robinson, we are unable to print here the usual Editor's Preface. We should therefore like to address a brief word to the readers.

It was Professor Robinson's intention to write, with the assistance of a number of co-workers, a comprehensive account of solar radiation, covering the theoretical and experimental information relating to solar radiation in the broadest sense, the fundamental aspects of the applications of solar energy, and practical information on radiation measurements.

In October 1964 Professor Robinson died suddenly, his task of editing the work still unfinished. But thanks to the generous cooperation of various friends and colleagues, who prefer to remain anonymous, it was possible to complete the editing of the manuscripts. We hope that the final product corresponds closely to the book he visualized and that it may serve as a tribute to his life's work.

LIST OF CONTRIBUTORS

The following contributors have collaborated in writing various sections of '*Solar Radiation*'.

U. BECKER, Freiburg i.Br., Germany
(part of Chapter 2).

M. Y. BEN-SIRA, Department of Physics, Technion – Israel Institute of Technology, Haifa, Israel
(Chapter 8, Section 2).

S. FRITZ, U.S. Weather Bureau, Washington, D.C., U.S.A.
(part of Chapter 3).

F. MÖLLER, Meteorological Institute of the University, Munich, Germany
(part of Chapter 3).

W. SCHÜEPP, Meteorological Service, Leopoldville, Congo (present address: Meteorological Observatory, Basle, Switzerland)
(Chapter 4).

G. YAMAMOTO, Geophysical Institute, Tohoku University, Sendai, Japan
(part of Chapter 3).

All the remaining material was contributed by the Editor.

CONTENTS

Chapter 7 – INSTRUMENTS AND EXPERIMENTAL METHODS

Chapter 1

THE RADIATION EMITTED BY THE SUN

1. LIMITS OF THE SOLAR SPECTRUM

The electromagnetic spectrum of the sun may be studied by various methods: by extrapolation from spectral measurements at various heights above sea level (including extra-atmospheric rocket measurements), partly by indirect methods, for example, by examining the state of the ozone layer in the upper atmosphere and of the various layers of the ionosphere [1–3], and by direct radio reception from the sun [4–6].

The electromagnetic spectrum emitted by the sun extends from fractions of an Ångström to hundreds of metres. It exhibits several gaps which should not, however, be confused with absorption lines due to the terrestrial and solar atmospheres, e.g. the Fraunhofer lines. The solar spectrum is usually divided into the wavelength regions indicated in Table 1.1.

The spectrum radiated by the so-called quiet sun in the region between 2500 and 30,000 Å carries 98 % of the total emitted energy. The solar radiation outside these limits is important but is of a lower intensity (see Table 1.2).

At wavelengths greater than 30,000 Å in the infrared range, almost all the energy is absorbed by atmospheric water vapour and carbon dioxide. In the ultraviolet there is a limiting wavelength of approximately 2860 Å for

TABLE 1.1. Classification of solar radiation according to wavelength.
($1 \text{ cm} = 10^8 \text{ Å} = 10^4 \mu$)

< 10 Å	X rays and γ rays
10 Å – 2000 Å	Far ultraviolet
2000 Å – 3150 Å	Middle ultraviolet
3150 Å – 3800 Å	Near ultraviolet
3800 Å – 7200 Å	Visible
7200 Å – 1.5 μ	Near infrared
1.5 μ – 5.6 μ	Middle infrared
5.6 μ – 1000 μ	Far infrared
> 1000 μ	Micro- and radio-waves

TABLE 1.2. Solar spectral irradiance data*.

Wavelength λ is in microns, the mean zero air mass spectral irradiance H_λ is in watts $\cdot$ cm^{-2} $\cdot$ micron^{-1}, and P_λ is the percentage of the solar constant associated with wavelengths shorter than wavelength λ.

λ	H_λ	P_λ	λ	H_λ	P_λ	λ	H_λ	P_λ	λ	H_λ	P_λ
0.22	0.0030	0.02	0.395	0.120	8.54	0.57	0.187	33.2	1.9	0.01274	93.02
0.225	0.0042	0.03	0.40	0.154	9.03	0.575	0.187	33.9	2.0	0.01079	93.87
0.23	0.0052	0.05	0.405	0.188	9.65	0.58	0.187	34.5	2.1	0.00917	94.58
0.235	0.0054	0.07	0.41	0.194	10.3	0.585	0.185	35.2	2.2	0.00785	95.20
0.24	0.0058	0.09	0.415	0.192	11.0	0.59	0.184	35.9	2.3	0.00676	95.71
0.245	0.0064	0.11	0.42	0.192	11.7	0.595	0.183	36.5	2.4	0.00585	96.18
0.25	0.0064	0.13	0.425	0.189	12.4	0.60	0.181	37.2	2.5	0.00509	96.57
0.255	0.010	0.16	0.43	0.178	13.0	0.61	0.177	38.4	2.6	0.00445	96.90
0.26	0.013	0.20	0.435	0.182	13.7	0.62	0.174	39.7	2.7	0.00390	97.21
0.265	0.020	0.27	0.44	0.203	14.4	0.63	0.170	40.9	2.8	0.00343	97.47
0.27	0.025	0.34	0.445	0.215	15.1	0.64	0.166	42.1	2.9	0.00303	97.72
0.275	0.022	0.43	0.45	0.220	15.9	0.65	0.162	43.3	3.0	0.00268	97.90
0.28	0.024	0.51	0.455	0.219	16.7	0.66	0.159	44.5	3.1	0.00230	98.08
0.285	0.034	0.62	0.46	0.216	17.5	0.67	0.155	45.6	3.2	0.00214	98.24
0.29	0.052	0.77	0.465	0.215	18.2	0.68	0.151	46.7	3.3	0.00191	98.39
0.295	0.063	0.98	0.47	0.217	19.0	0.69	0.148	47.8	3.4	0.00171	98.52
0.30	0.061	1.23	0.475	0.220	19.8	0.70	0.144	48.8	3.5	0.00153	98.63
0.305	0.067	1.43	0.48	0.216	20.6	0.71	0.141	49.8	3.6	0.00139	98.74
0.31	0.076	1.69	0.485	0.203	21.3	0.72	0.137	50.8	3.7	0.00125	98.83
0.315	0.082	1.97	0.49	0.199	22.0	0.73	0.134	51.8	3.8	0.00114	98.91
0.32	0.085	2.26	0.495	0.204	22.8	0.74	0.130	52.7	3.9	0.00103	98.99
0.325	0.102	2.60	0.50	0.198	23.5	0.75	0.127	53.7	4.0	0.00095	99.05
0.33	0.115	3.02	0.505	0.197	24.2	0.80	0.1127	57.9	4.1	0.00087	99.13
0.335	0.111	3.40	0.51	0.196	24.9	0.85	0.1003	61.7	4.2	0.00080	99.18
0.34	0.111	3.80	0.515	0.189	25.6	0.90	0.895	65.1	4.3	0.00073	99.23
0.345	0.117	4.21	0.52	0.187	26.3	0.95	0.0803	68.1	4.4	0.00067	99.29
0.35	0.118	4.63	0.525	0.192	26.9	1.0	0.0725	70.9	4.5	0.00061	99.33
0.355	0.116	5.04	0.53	0.195	27.6	1.1	0.0606	75.7	4.6	0.00056	99.38
0.36	0.116	5.47	0.535	0.197	28.3	1.2	0.0501	79.6	4.7	0.00051	99.41
0.365	0.129	5.89	0.54	0.198	29.0	1.3	0.0406	82.9	4.8	0.00048	99.45
0.37	0.133	6.36	0.545	0.198	29.8	1.4	0.0328	85.5	4.9	0.00044	99.48
0.375	0.132	6.84	0.55	0.195	30.5	1.5	0.0267	87.6	5.0	0.00042	99.51
0.38	0.123	7.29	0.555	0.192	31.2	1.6	0.0220	89.4	6.0	0.00021	99.74
0.385	0.115	7.72	0.56	0.190	31.8	1.7	0.0182	90.83	7.0	0.00012	99.86
0.39	0.112	8.13	0.565	0.189	32.5	1.8	0.0152	92.03			

* Private communication from F. S. Johnson, Lockheed Aircraft Corporation, Sunnyvale, Calif.

radiation reaching sea level, whilst still shorter wavelengths are absorbed by the ozone layer in the upper atmosphere (cf. Chapter 3).

Fig. 1.1 shows a plot of the data given in Table 1.2 together with a spectrum obtained earlier by P. Moon at the Smithsonian Institution.

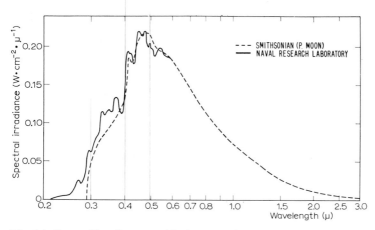

Fig. 1.1. Spectral irradiance outside the atmosphere.

Some excellent photographs were obtained in the far ultraviolet (170–2100 Å) by R. Tousey *et al.* of the U.S. Naval Research Laboratory, Washington, D.C. The spectra were recorded at altitudes between 125 and 220 km (March 1959, April 1960, June 1961). The far ultraviolet spectrum of the sun shown in Fig. 1.2 was obtained by W. E. Austin, J. D. Purcell and R. Tousey [7] of the U.S. Naval Research Laboratory, on June 21, 1961. A grazing incidence spectrograph with a 40 cm radius grating was used with a thin aluminium filter over the slit. The exposure was 112 sec and the maximum altitude was 197 km. Nearly all the lines between 310 and 700 Å are higher order images of intense first-order lines which were not photo-

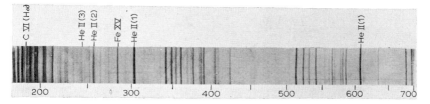

Fig. 1.2. Far-ultraviolet spectrum of the sun.

References pp. 26–28

graphed above 310 Å because of the strong absorption by atmospheric molecular nitrogen. Most of the lines are unidentified, but it appears probable that the Hα line of C VI and the resonance line of Fe XI are present.

Because of the limited scope of this book we shall confine our attention to that part of the solar spectrum which is usually measured. For literature dealing with shorter and longer wavelengths, see [8–10].

2. COMPARISON OF SOLAR AND BLACK-BODY RADIATION

In the region of the solar spectrum with which we shall be concerned in this book, the sun may be regarded as a 'temperature radiator' or 'heat radiator', *i.e.* a body which emits radiant energy by virtue of its temperature. It also converts absorbed radiation into heat. The intensity and spectral distribution of the emitted radiation depend on the temperature of the radiating body and on its nature.

Let us briefly review the laws of radiation. The first of these is Kirchhoff's law which states that the ratio of the emitted energy to the absorbed energy is solely a function of temperature T and the wavelength λ. Thus

$$E_{em}/E_{ab} = f(\lambda, T) \tag{1.1}$$

where E_{em} is the amount of energy emitted per unit area of the body and E_{ab} is the fraction of incident energy which is absorbed by the body, *i.e.* neither reflected nor transmitted. When $E_{ab} = 1$ the body is referred to as perfectly 'black'. A black body absorbs all the energy incident upon it. It follows from Eq. (1.1) that a black body also possesses the highest emissivity at all wavelengths as compared with all other bodies. The radiation emitted by it is termed black-body radiation.

The second, Planck's law is concerned with the spectral energy distribution of black-body radiation. According to this law

$$E_{\lambda} d\lambda = \frac{(hc^2/\lambda^5) \cos \theta \, d\lambda}{\exp(hc/k\lambda T) - 1} \tag{1.2}$$

where $E_{\lambda} d\lambda$ is the amount of linearly polarized energy emitted by a unit area in the wavelength region $\lambda \rightarrow \lambda + d\lambda$ per second per unit solid angle, by a black body in thermal equilibrium with its surroundings, θ is the angle with the normal to the emitting area, h is Planck's constant $(6.626 \times 10^{-27}$

erg · sec), k is Boltzmann's constant (1.3807×10^{-16} erg · deg^{-1}) and c is the velocity of light. Fig. 1.3 shows three curves calculated from Eq. (1.2) for $T = 7000$, 6000 and 5000°K respectively.

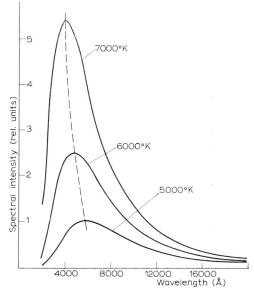

Fig. 1.3. Spectral curves for a black body.

A further relationship which can be derived from Planck's law is the Wien displacement law:

$$\lambda_m T = 0.288 \text{ cm} \cdot \text{deg} \tag{1.3}$$

where λ_m is the wavelength corresponding to maximum energy in the spectrum of the black body at an absolute temperature T.

The total energy emitted per unit area per second by a black body is given by the Stefan–Boltzmann law:

$$E_{em} = \sigma T^4 \tag{1.4}$$

where $\sigma = 5.73 \times 10^{-5}$ erg · cm^{-2} · deg^{-4} · sec^{-1}.

The temperature of a black body may be determined from any one of these laws.

A body emitting radiation whose intensity is lower than that of a black body, by the same factor at all wavelengths, is referred to as 'grey'. The

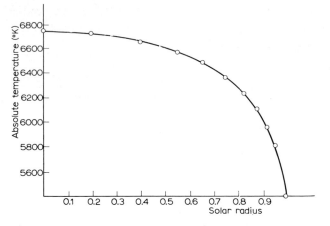

Fig. 1.4. Temperature variation across the solar disc as seen by a terrestrial observer [10].

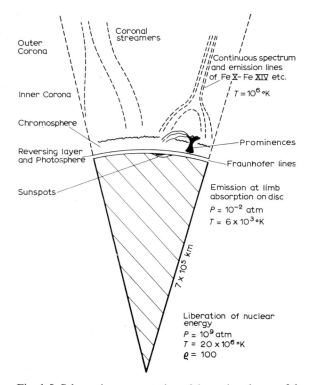

Fig. 1.5. Schematic representation of the various layers of the sun [1].

temperature deduced for a grey body from the Stefan–Boltzmann law is lower than the 'real' temperature. The temperature determined in this way is called black, effective or radiation temperature. Another quantity, which is used for celestial bodies, is the colour temperature which is defined as the temperature of a black body with the same relative energy distribution in its spectrum as in the spectrum of the given body.

Some difficulties are encountered in the determination of the temperature of the sun from the above laws of radiation.

(a) Different spectral regions correspond to different temperatures. For example, in the region near 24,000 Å the sun radiates as a grey body at 6000°K with an emissivity of 0.999. Between 1000 and 2000 Å it behaves as a black body at 4500°K, and between 40 and 100 Å the radiation corresponds to a black body at 5×10^5°K. Flares emit radiation of approximately 3 Å as a black body at 4×10^6°K.

(b) There is a temperature variation across the solar disc as seen from the earth (Fig. 1.4).

(c) Solar radiation is emitted simultaneously by layers of varying depths, temperature and physical properties. Fig. 1.5 shows a schematic representation of the solar layers and their temperatures.

The photosphere, the reversing layer and the chromosphere are of particular interest in connection with the radiation discussed in this book (ultraviolet, visible and near infrared). The photosphere is the deepest of the three and is the main source of solar radiation. It has a density of about 10^{-8} g $\cdot$ cm^{-3} and emits a continuous spectrum. Table 1.3 gives a summary of various determinations of the temperature in the photosphere.

The photosphere is not of uniform brightness and exhibits regions which are brighter or darker than the background. The brighter parts include

TABLE 1.3. Photospheric temperature determinations [10].

Method	Region Å	Temperature (°K)
Colour temperature:		
Mulders	3000–4000	4850
Mulders	4100–9500	7140
Arnulf et al.	3000–4316	6200
Effective radiation	Total	5770
Visible radiation (Brill)	Total	6050
Wien's law	Total	6080

the so-called granules which appear 35–40% brighter than the background when photographed from the earth's surface. They cover about 35% of the surface of the photosphere and number 2.5×10^6. The mean distance between the centres of adjacent granules is $d = 2.5''–2.6''$. For $d = 2.5''$ the peak of intensity is at 4650 Å and for $d = 2.6''$ it occurs at 6000 Å. The granules are unstable and have lifetimes of the order of a few minutes. Their temperature differs from the background temperature by approximately 175°C when measured at sea level. The diameter of the granules is 400–1000 km [11–15].

More recent measurements [26] from a balloon at 28 km showed that the mean life of the granules is 8.6 ± 0.2 min and the r.m.s. difference between their temperature and the background temperature is $\pm 92°$K. The results of these measurements are more reliable because distortions in the shape and brightness, caused by the atmosphere, were avoided in this case.

Another type of brighter specks are the faculae which are on the average brighter than the photospheric background by a factor of 1.1. Fig. 1.6 shows granules, small dark areas called pores, and bigger dark areas (small sunspots). Fig. 1.7 shows sunspots and faculae.

There are two types of faculae, namely, those visible in monochromatic light (*e.g.* the Hα radiation) over the entire solar disc, and secondly, very weak faculae close to the solar limb and visible in the continuum. The former belong to the chromosphere and the latter to the photosphere. They are practically the same but are observed at different depths. Their lifetime is about 30 minutes. They are small, bright formations, becoming weaker and larger towards the end of their existence. The pores and sunspots belong to the darker parts of the photosphere. The temperature of sunspots is 4300–4500°K and they appear as dark spots on a bright background. The number and the dimensions of the pores and sunspots change continuously and this adds to the difficulties enumerated above [18–22].

A further difficulty in the determination of the temperature of the photosphere is its thickness. Although the latter is comparatively small (about 300 km) radiation coming from other regions is partly absorbed by it. Finally, the photosphere is not in thermal equilibrium, assumed in the derivation of the radiation laws.

The next layer considered lies immediately above the photosphere [23–24] and is termed the 'reversing layer', because it is cooler (about 5300°K) than the photosphere and consequently gives rise to absorption lines in the spectrum.

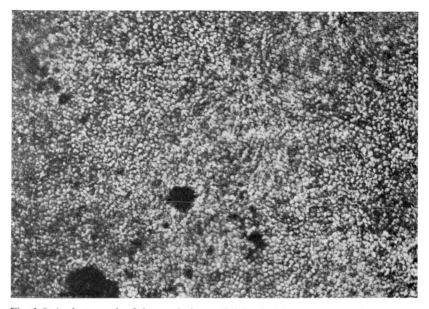

Fig. 1.6. A photograph of the sun in integral light showing granules, pores and sunspots [6].

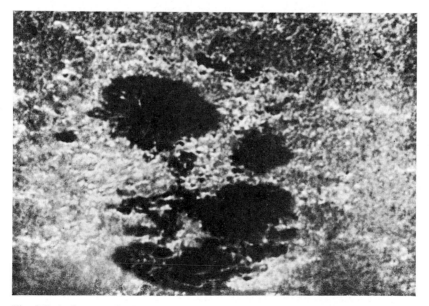

Fig. 1.7. A photograph of the sun showing sunspots and faculae [6].

The third layer which we shall briefly discuss is the chromosphere, or the 'colour sphere', so called because of its particular colour (red). It is an extension of the reversing layer, without a clearly defined boundary between

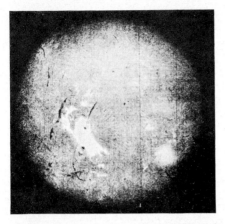

Fig. 1.8. A photograph of the sun in Hα light (Meudon, July 1946).

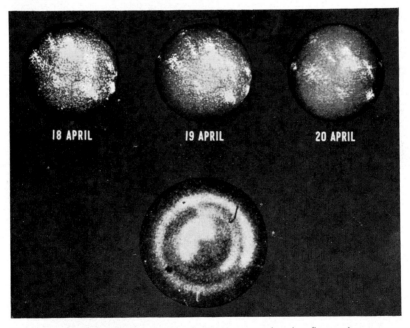

Fig. 1.9. A photograph of the sun in the X-ray spectral region (bottom) as compared with photographs in Ca–K light.

them, and appears as an irregular disc during solar eclipses. Because of the prevalence of hydrogen, it emits strongly in the Hα line at 6563 Å. The intensity of the Hα line decreases up to the height of 12,000 km above the photosphere. The H and K lines of ionized calcium appear up to the height of 14,000 km above the photosphere. The pressure decreases and the temperature increases in the outward direction.

The chromosphere, too, is not uniformly bright. When observed with spectroheliographs and interference filters it exhibits both bright and dark areas. Fig. 1.8 shows a photograph of the sun in Hα light. The large bright

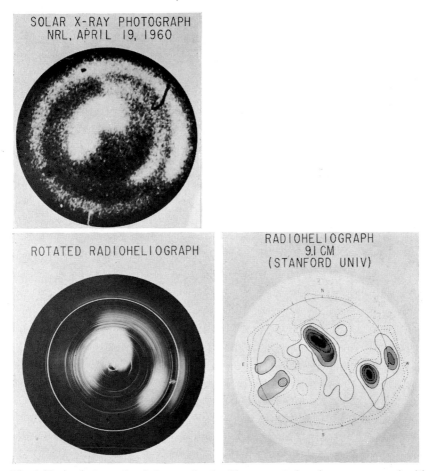

Fig. 1.10. A photograph of the sun in the X-ray spectral region as compared with a rotated radio-heliograph of $\lambda = 9.1$ cm.

References pp. 26–28

area is an eruption called a flare; the small bright specks are chromospheric faculae. The dark specks, called filaments, are prominences above the disc [25, 27–29].

More recently, the sun was photographed in the X-ray spectral region by means of a pin-hole camera by H. Friedman [26] of the U.S. Naval Research Laboratory, Washington, D.C.; the hole was covered by a material which cut off all radiation except the X rays. The camera was carried by a rocket which spun on its axis and so produced a smeared-out effect in the photograph. In Fig. 1.9 this photograph is compared with three photographs taken in Ca–K light. In order to investigate this smearing effect, a radio-heliograph of $\lambda = 9.1$ cm (Fig. 1.10) was spun in the same way as the rocket which carried the camera for the X-ray photograph; the result is given in the lower left-hand part of Fig. 1.10. The upper part of Fig. 1.10 shows the enlarged X-ray photograph for comparison.

We may conclude that a simple determination of the temperature of the sun is impossible and that the energy distribution in the solar spectrum cannot be expected to correspond to that of a black body.

3. THE TEMPERATURE OF THE SOLAR DISC

Let us now regard the sun as a uniformly radiating body and compare its spectral curve with black-body energy distributions at various temperatures. The absorption lines in the solar spectrum will be neglected, *i.e.* the spectral curve will be smoothed out.

Fig. 1.11 shows three black-body energy distributions (5000, 6000 and 7000°K) and the smoothed solar curve. The shaded area corresponds to the visible part of the spectrum (not as seen by the human eye, but the total radiation in this range). Although the solar radiation curve is close to the black-body distribution for $T = 6000$°K, there are important differences between them. Firstly, the intensity maximum is not at 4800 Å as predicted by Eq. (1.3) but at 4700 Å. Secondly, the spectrum is cut off on the ultraviolet side, and lies below the 6000°K curve in the infrared.

Fig. 1.12 shows a comparison of the ultraviolet part of the black-body spectrum with the solar spectrum recorded at two heights above the earth's surface.

The temperature of the sun as determined from the earth's surface decreases from the centre towards the limb. This effect is called limb darken-

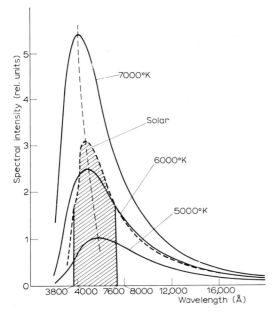

Fig. 1.11. Spectral curve of the sun compared with black-body distributions [4].

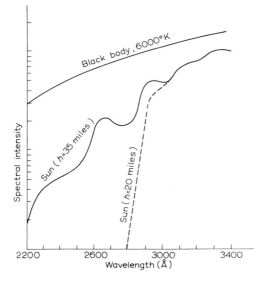

Fig. 1.12. Spectral curve of the sun compared with the black-body distribution at 6000°K (ultraviolet region) [10].

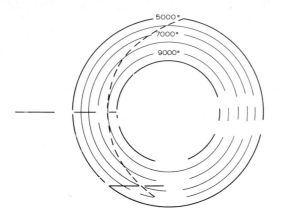

Fig. 1.13. Origin of limb darkening [57].

ing, and is illustrated in Fig. 1.4. The reason for this variation will be clear from an inspection of Fig. 1.13. Thus, the layers seen at the centre of the disc are deeper (hotter) than those near the limb.

Furthermore, the effect is selective and depends on the wavelength (Fig. 1.14). Thus, in the red part of the spectrum the darkening is about 1%, while in the green and the ultraviolet it is approximately 8% and 40% respectively.

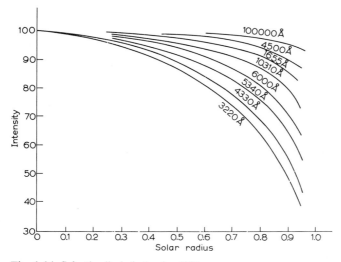

Fig. 1.14. Selective limb darkening [10].

4. THE SOLAR CONSTANT

The amount of energy passing per unit time through a unit area at right angles to the direction of the solar beam outside the atmosphere must be known before one can compute the amount of energy emitted by the sun in all directions and in the direction of the earth. The concept of a 'solar constant' I_0 was introduced by A. Pouillet in 1837 in order to facilitate such calculations. A method for the determination of this quantity was first given by Langley in 1881.

The solar constant is defined as the quantity of solar energy (cal · cm^{-2} · min^{-1}) at normal incidence outside the atmosphere at the mean sun–earth distance (for particulars of this distance see Chapter 2, Section 1, p. 29) [30–37]. According to this definition it is useful to introduce a new unit of energy which is known as the Langley and is equivalent to 1 cal · cm^{-2}.

It is clear from the definition of I_0 that it must be computed by extrapolation from measurements at various heights h, or at various altitudes of the sun at a given place. Table 1.4 gives a list of determinations of the energy flux at various heights above sea level.

TABLE 1.4. I_0 values determined at different heights above sea level [10].

Place	Height above sea level (metres)	Energy (mean) (cal · cm^{-2} · sec^{-1})
Sea level	0	1.40
Davos	1580	1.59
Jungfraujoch	3400	1.63
Mount Whitney	4350	1.72
Balloon	19500	1.89

A precise value of the solar constant cannot be obtained by a strict extrapolation from these results because the various strata of the earth's atmosphere have different transparencies. A simple extrapolation from measurements at various altitudes of the sun at the same point is again not possible, because the total atmospheric extinction coefficient is a function of wavelength (cf. Chapter 3).

Fig. 1.15 shows some of the experimental results obtained by F. W. Goetz at Arosa, Switzerland, at various altitudes of the sun. The curves marked A, B and C show the integrated flux, the flux for $\lambda > 6000$ Å and the flux for $\lambda > 600$ Å respectively.

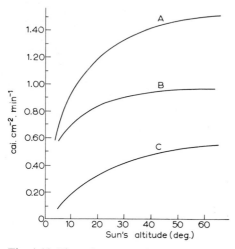

Fig. 1.15. The solar energy flux as a function of the sun's altitude for different spectral regions [10].

Extrapolation from measurements at various altitudes of the sun can be carried out with the aid of the Bouguer-Lambert law:

$$E_\lambda = E_{0\lambda}(q_\lambda)^l \tag{1.5}$$

where $E_{0\lambda}$ is the initial energy, E_λ is the reduced energy, l is the path length in the atmosphere, and q_λ is the proportion of energy transmitted by a layer of unit thickness (all at given λ). Fig. 1.16 shows the dependence of q_λ on λ. It must be concluded from this plot that the extrapolation to the energy distribution in the spectrum outside the atmosphere has to be carried out separately for each wavelength. The integrated area below the extrapolated curve gives the total energy emitted by the sun within the given wavelength range.

The Fraunhofer lines have no effect on such an extrapolation because they are due to the solar atmosphere and therefore do not change with the altitude γ of the sun. It will be shown later that in order to determine the solar constant from measurements at different γ, it is necessary to determine both the total energy reaching the earth and the energy in the various spectral regions.

The extraterrestrial radiation energy E_0 (at the top of the earth's atmosphere), which is included in the wavelength range $\lambda_1 \rightarrow \lambda_2$ is given by

$$E_0 = \int_{\lambda_1}^{\lambda_2} E_{0\lambda} \, d\lambda \tag{1.6}$$

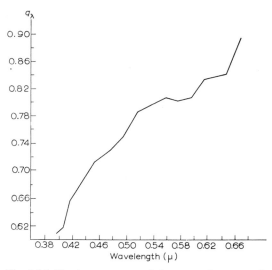

Fig. 1.16. The transparency of the atmosphere as a function of wavelength [10].

where $E_{0\lambda}$ is the specific energy for the given λ, or for a narrow region centred on this value of the wavelength. Since the radiation outside the atmosphere cannot be measured (except by satellites), an extrapolation from terrestrial measurements is usually employed. For the radiation energy on the earth, Eq. (1.6) will be rewritten in the form

$$E = \int_{\lambda_1}^{\lambda_2} E_\lambda \, d\lambda \tag{1.7}$$

We shall now find the relation between $E_{0\lambda}$ and E_λ. The attenuation of radiation of intensity I_λ on passing through the atmosphere is described by

$$dI_\lambda = -I_\lambda a_\lambda \, dl = -I_\lambda a_\lambda \, d(h \sec z) \tag{1.8}$$

where l is the path length in the atmosphere, z is the zenith distance of the sun (cf. Chapter 2, Section 2, p. 31) and a_λ is the extinction coefficient at the given wavelength λ. In order to obtain the extinction due to the total atmosphere, an equivalent homogeneous atmosphere of height H is used. This atmosphere is of constant density which is equal to that of air under normal conditions (sea level, $0°C$, 760 mm Hg). For this atmosphere Eq. (1.8) yields

$$I_\lambda = I_{0\lambda} \exp(-a_\lambda H \sec z) \tag{1.9}$$

where $I_{0\lambda}$ is the intensity outside the atmosphere at the given wavelength λ.

This intensity can be determined from a minimum of two measurements of I_λ for the same λ and different z, provided it is assumed that a_λ remains constant. The required relation between E_λ and $E_{0\lambda}$ is

$$E_\lambda = E_{0\lambda}(I_\lambda/I_{0\lambda}) \tag{1.10}$$

This relation will, of course, hold both in absolute and relative units. Thus, the total intensity I reaching the earth is given by

$$I = I_0 \frac{\displaystyle\int_{\lambda_1}^{\lambda_2} E_\lambda \mathrm{d}\lambda}{\displaystyle\int_{\lambda_1}^{\lambda_2} E_{0\lambda} \mathrm{d}\lambda} = I_0 \frac{\displaystyle\int_{\lambda_1}^{\lambda_2} E_{0\lambda}(I_\lambda/I_{0\lambda})\mathrm{d}\lambda}{\displaystyle\int_{\lambda_1}^{\lambda_2} E_{0\lambda} \mathrm{d}\lambda} \tag{1.11}$$

In this equation I and I_λ are measured directly (cf. Chapter 7) while $E_{0\lambda}$ and $I_{0\lambda}$ are obtained by extrapolation.

However, I_0 cannot be computed exactly from Eq. (1.11) for two reasons. Firstly, the limits of the spectrum outside the atmosphere and on the earth are different because of the atmospheric cut-off in the ultraviolet and the infrared. Secondly, the measurements depend on the method and instruments employed as well as on the adopted scale (see Chapter 7).

Two methods have been developed at the Smithsonian Institution for the determination of the solar constant, known respectively as the 'short' and the 'long' method.

In the long method a spectrobolometer is used to determine the relative intensity in 40 narrow spectral regions of equal width in the range 3460–24,400 Å, and magnitude of E_λ for different zenith distances. The extraterrestrial values of these intensities $E_{0\lambda}$ are determined in relative units by extrapolation. The values of E_λ and $E_{0\lambda}$ are then plotted against λ and the smoothed curves which are obtained as a result are integrated. This yields the ratio of integrals in Eq. (1.11) in the range 3460–24,400 Å. The spectrobolometric measurements are carried out in parallel with determinations of the total intensity using an absolute standard pyrheliometer (see Chapter 7) for the same zenith distance. The results obtained in this way are inserted into Eq. (1.11) and I_0 is evaluated. However, the resulting values of the solar constant must be corrected for atmospheric losses in the ultraviolet and the infrared within the wavelength limits of the spectrobolometer [38,39] and also for instrumental losses. Finally, the total area of the Fraunhofer lines (in the same units) must be subtracted from the integrals of Eq. (1.11). Fig. 1.17 shows a spectrobologram and the smoothed curve.

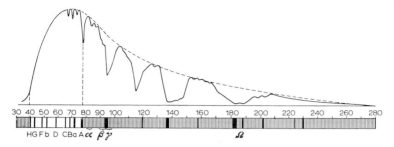

Fig. 1.17. A detailed solar spectrum (including Fraunhofer lines) and the smoothed curve [57].

The expression for the numerical determination of the solar constant may be written down in the form

$$I_0 = I \frac{\Sigma E_{0\lambda} + \Delta E_{0\lambda}(\text{UV}) + \Delta E_{0\lambda}(\text{IR})}{\Sigma E_\lambda + \Delta E_\lambda(\text{UV}) + \Delta E_\lambda(\text{IR})} \tag{1.12}$$

where $\Sigma E_{0\lambda}$ and ΣE_λ include corrections for the Fraunhofer lines in the given spectral region. The ultraviolet and infrared corrections $\Delta E_{0\lambda}(\text{UV})$ and $\Delta E_{0\lambda}(\text{IR})$ in the numerator are different from the corresponding corrections $\Delta E_\lambda(\text{UV})$ and $\Delta E_\lambda(\text{IR})$ in the denominator.

Thus, $\Delta E_{0\lambda}(\text{UV})$ is the correction for the range between 3400 Å and the minimum wavelength at which the intensity is still measurable outside the atmosphere, and $\Delta E_{0\lambda}(\text{IR})$ is the correction for the range between 24,400 Å and the maximum wavelength at which the intensity is still measurable outside the atmosphere. Finally, $\Delta E_\lambda(\text{UV})$ is the correction for the range between 3460 and 2960 Å (minimum wavelength on the earth's surface) and $\Delta E_\lambda(\text{IR})$ is the correction for the interval between 24,400 Å and the maximum measurable wavelength on the earth*.

It is evident from the above description of the long method that it is very complicated and time consuming. This affects its accuracy because of changes in the atmospheric transparency during the measurements, even under the favourable climatic conditions of Mount Wilson, Montezuma, and so on.

The absolute standard pyrheliometer used in the long method is very

---

* F. S. Johnson refers to the correction in the denominator as 'spectrobologram corrections' while those in the numerator he calls 'zero air-mass corrections' (private communication of March 30, 1959; see also a detailed discussion in Johnson's paper 'The Solar Constant' [32].

difficult to work with and requires long measuring periods, which adds to the difficulties of the methods [40–42].

Sub-standards are frequently used to reduce the experimental difficulties. This involves further uncertainties associated with the conversion factor scale from the standard to the sub-standard (this will be discussed in Chapter 7).

Before we proceed to the short method let us consider the ultraviolet and infrared corrections. The infrared correction can be determined by two methods. In the first method the energy distribution in the spectrum is measured with a rock-salt spectrograph, which can be done up to $\lambda = 100,000$ Å. The corrections are deduced from these measurements. Absorption of the infrared radiation by atmospheric water vapour has to be taken into consideration (see Chapter 3). This is done by determining the infrared spectrum for different amounts of water vapour in the atmosphere and then extrapolating to zero water content. This is essential as the correction is for I_0 outside the atmosphere.

. The second method of determining the infrared correction is to compute the missing infrared part of the spectrum from the Planck formula (Eq. (1.2)) for a black or grey body at 6000°K.

Because of the absorption of ultraviolet radiation by the ozone layer in the upper atmosphere and the dependence of the flux of this radiation on the transparency of the atmosphere (see Chapter 3), the ultraviolet correction is best obtained from rocket measurements.

The short method is based on only one measurement of the energy distribution in the solar spectrum on the earth at a given zenith distance, with subsequent extrapolation to the outside of the atmosphere. The extrapolation can only be carried out if the atmospheric attenuation is known for all the 40 spectral regions mentioned above. This attenuation is due to absorption (mainly by water vapour) and scattering. The latter may be determined from the intensity of the circumsolar radiation. There are empirical relations between the reduction in the intensity, the water vapour content and the intensity of the circumsolar radiation.

A single measurement in the 40 spectral regions at a given zenith distance is sufficient for the extrapolation to be carried out. Subsequent procedure is the same as in the long method. The time needed for the determination of the solar constant by the short method is about one-tenth of that for the long method. There are, however, some doubts as to the accuracy of the short method because simultaneous determinations at different places are

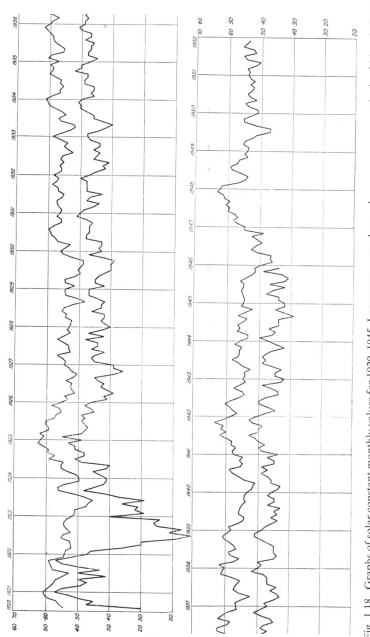

Fig. 1.18. Graphs of solar constant monthly values for 1920–1945. Lower curve: as observed; upper curve: a synthesis of 14 periodicities. Predictions 1920–1923 and 1946–1951 in upper curve. Ordinates in cal · cm^{-2} · min^{-1} when prefixed by 1.9 [3].

found to yield different values of I_0. Moreover, it seems that this method is not independent of weather conditions [43,44].

The following are some of the values of I_0 obtained by different workers: Smithsonian Institution: 1.982; Linke: 1.979; Nicolet: 1.981; Johnson: 2.00 $\pm$ 2%; Quirck: 1.99 and 2.02 (1950, Aerobee rocket at $h=70$ km); U.S. Weather Bureau: 2.02, 1.995 and 2.01 ($h=90{,}000$ ft, Smithsonian scale of 1913; cf. Chapter 7).

The differences between these values of the solar constant are larger than the differences between the values of I_0 obtained at a given point at different times. The accuracy of the constants is the same and is sufficient for the determination of even small variations in them. Fig. 1.18 shows a plot of the magnitude of I_0 over a number of years.

A detailed analysis of this variation in I_0 at various stations has shown that it is largely associated with weather elements [45] and the particular instruments employed. A correlation has been found between the solar-constant variation and sunspots. E. Petit supposed that if such a correlation exists, it should be more clearly defined in the ultraviolet than at longer wavelengths. He compared the regions near $\lambda=3200$ Å and $\lambda=5000$ Å and found a correlation for the years 1924–1931. In 1953 L. B. Aldrich and W. H. Hoover published their results on the correlation between sunspots and the solar-constant variation. They found an increase of 0.6% during the sunspot maximum of 1948 as compared with the minimum of 1944.

The fact that the variations in the solar constant depend on the accuracy of the instruments employed in its determination is confirmed by the reduction in the deviations of the diurnal values of I_0 from the mean when the accuracy is increased. The problem of the periodicity of the long-range variation in I_0 has not been solved either. In May 1959 Johnson and Iriarte [46] reported some results on the variation in the solar constant during the preceding six years. They measured the light reflected from Uranus and Neptune and compared it with direct light from nearby stars. Preliminary results have indicated an increase of about 2% during the six year period.

The uncertainties associated with determinations of the solar constant and its variation and periodicity have been widely discussed in the literature [47–56].

5. TOTAL ENERGY EMITTED BY THE SUN

It is evident from the definition of the solar constant that the mean value of the energy emitted by the sun per minute is given by

$$E_s = 4\pi R_0^2 I_0 \tag{1.13}$$

where R_0 is the mean earth–sun distance. The power emitted by the sun can be calculated from this expression and turns out to be 3.85×10^{23} kW or 5.16×10^{23} hp. The part of this energy intercepted by the earth is given by

$$E_e = \pi r_0^2 I_0 \tag{1.14}$$

where r_0 is the mean radius of the earth. The associated power is 1.79×10^{14} kW or 2.40×10^{14} hp.

6. RECENT THEORIES OF THE SOURCE OF SOLAR ENERGY

The old theories of combustion as the source of solar energy, Helmholtz's contraction hypothesis, natural radioactive decay, and so on have been found to be inadequate. Such sources do not produce sufficient energy and become exhausted within a much shorter time than required. Moreover, they are not consistent with the present knowledge of the chemical constitution and physical state of the sun.

The recent theory of Bethe leads to an agreement with the temperature conditions on the sun, the amount of liberated energy, the source lifetime and the amount of helium in the sun. According to this theory the energy liberating process is a thermonuclear one, leading to the production of alpha particles (helium nuclei) with carbon and nitrogen as catalysts. Starting, for example, with $^{12}_{6}C$ the process is as follows. A collision with a proton ($^{1}_{1}H$) leads to the formation of the nitrogen isotope $^{13}_{7}N$. The mass difference between $^{12}_{6}C + ^{1}_{1}H$ and $^{13}_{7}N$ is transformed into energy in accordance with Einstein's law

$$E = mc^2 \tag{1.15}$$

where m is the mass in grams, c is the velocity of light (3×10^{10} cm · sec^{-1}) and E is the energy in ergs (process 1 in Fig. 1.19).

The isotope $^{13}_{7}N$ is unstable and decays into the carbon isotope $^{13}_{6}C$ with the emission of a positron (process 2 in Fig. 1.19). This isotope collides

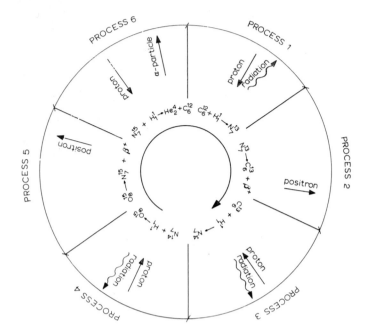

Fig. 1.19. Schematic representation of nuclear reactions responsible for energy liberation on the sun.

with a further proton and becomes converted into ordinary nitrogen $^{14}_7\text{N}$ and the mass difference is again radiated in the form of gamma rays (process 3). The $^{14}_7\text{N}$ collides with a third proton forming the unstable oxygen isotope $^{15}_8\text{O}$ and the mass difference is converted into gamma rays (process 4). The unstable isotope decays to $^{15}_7\text{N}$ with the emission of a positron (process 5). Finally, the $^{15}_7\text{N}$ collides with a fourth proton and disintegrates into $^{12}_6\text{C}$ and an alpha particle (process 6).

At the end of the chain one alpha particle is formed as a result of four proton collisions but the amounts of carbon and nitrogen remain unchanged. The duration of such a chain is 5×10^6 years. Since the mass of one proton is 1.672×10^{-24} g and the mass of one alpha particle is 6.644×10^{-24} g, it follows that the mass loss entailed in the creation of one alpha particle from four protons is 0.044×10^{-24} g. According to Einstein's law (Eq. (1.15)), one gram is equivalent to 9×10^{20} ergs and hence one gram–proton liberates 6.3×10^{18} ergs when converted into alpha particles. Since about 0.007 of the mass of the four protons is lost in the creation of an alpha particle, it follows that only 0.007 of the total mass of protons in the sun is converted into radiation.

A further confirmation of the validity of the above mechanism is the fact that the mass of helium now present in the sun is approximately 2% of the total mass. This is in good agreement with Bethe's theory provided that there were no alpha particles in the sun at the beginning and if it has been radiating for about 2×10^9 years. This figure is the lower limit of the age of the sun as deduced from meteors by the radioactive method. The figure of 2% is calculated from the amount of energy emitted during this period $(2.4 \times 10^{50}$ ergs) and the mass loss during the creation of one alpha particle from four protons.

A knowledge of its store of energy, or the number of protons it contains, is necessary before one can determine the length of time for which the sun will continue to radiate with the present intensity. This has been computed by various workers, the results varying between 51% and 69% of the total mass. Assuming that 70% of the total mass of the sun is made up of protons and that the rate at which radiant energy is emitted remains constant, it may be shown that the sun will continue to emit radiation for another 10^{11} years.

The short-wave gamma radiation is degraded in energy on the way from the centre of the sun towards the surface. The wavelength continuously changes until at the surface most of the radiation is in the visible range.

Another possible mechanism which may act as the source of solar energy is the proton–proton reaction put forward by Bethe, Critchfield, Fowler and Lauritsen. According to this idea, a contraction process, which lasts for 5×10^7 years, takes place before a thermonuclear reaction becomes possible. The onset of this reaction leads to a release of energy. This counteracts the contraction process which would otherwise cause the collapse of the sun. The reaction can be tabulated as follows:

	reaction	mean lifetime
(a)	$2\,{}^1_1H + 2\,{}^1_1H \rightarrow 2\,D + 2\beta^+ + $ neutrino	10^{10} years
(b)	$2\,D + 2\,{}^1_1H \rightarrow 2\,{}^3_2He + \gamma$	10 sec
(c)	${}^3_2He + {}^3_2He \rightarrow {}^4_2He + 2\,{}^1_1H$	2×10^6 years

The mean half-life of a proton under the conditions which are at present accepted as existing in the interior of the sun is about 10^{10} years. This is a consequence of the small cross-section (about 10^{-24} cm^2) and the scarceness of targets in spite of the comparatively high density in the solar interior. It is also due to the fact that the colliding protons have electric charges of the same sign, which gives rise to repulsive forces between them. Finally, one must remember that the rate of successful proton–proton collisions is proportional to T^4 so that the main part of such collisions is restricted to the central part of the sun. The protons are more likely to collide than the other elements because the repulsive forces between the heavy-element nuclei are larger.

The positrons from reaction (a) annihilate themselves with the electrons while the neutrinos penetrate into the surrounding space. The two protons in reaction (c) are free to take part in a further reaction of type (a).

Calculations of changes in the mass and liberated energy show that the sun loses 4×10^6 ton $\cdot$ sec^{-1} or 0.5×10^{-21} of its total mass per second. This mass defect derives from the fact that $4\,H = 4.03257$ amu, ${}_2^4He = 4.00389$ amu ($\Delta m = 0.02868$ amu in every reaction).

REFERENCES

1. M. A. ELLISON, *The Sun and its Influence*, Routledge and Kegan Paul, London (1956), pp. 180–199.
2. M. WALDMEIER, *Ergebnisse und Probleme der Sonnenforschung*, Akad. Verlagsges., Leipzig (1955), pp. 8–11; *Astron. Mitt. Eidgenöss. Sternwarte Zürich*, No. 233 (1961), pp. 1–4.
3. C. DE JAGER, *Handbuch der Physik (Encyclopaedia of Physics)*, Springer, Berlin, Vol. 52 (1959), pp. 342–362.
4. C. DE JAGER, *Handbuch der Physik (Encyclopaedia of Physics)*, Springer, Berlin, Vol. 52 (1959), pp. 143–144, 250, 283–289, 296–308.
5. M. WALDMEIER, *Ergebnisse und Probleme der Sonnenforschung*, Akad. Verlagsges., Leipzig (1955), pp. 327–350.
6. J. L. LAWSEY AND S. F. SMERD, *The Sun* (edited by G. P. Kuiper), Univ. of Chicago Press, Chicago, Ill. (1953), pp. 466–531.
7. R. TOUSEY, Communication to the IAU at Berkeley, Calif. on August 16, 1961.
8. C. DE JAGER, *Handbuch der Physik (Encyclopaedia of Physics)*, Springer, Berlin, Vol. 52 (1959), pp. 208, 256, 261–264, 278.
9. R. TOUSEY, *The Sun* (edited by G. P. Kuiper), Univ. of Chicago Press, Chicago, Ill. (1953), pp. 667–668.
10. G. ABETTI, *The Sun*, Faber and Faber, London (1955), pp. 97, 259, 277.
11. G. ABETTI, *The Sun*, Faber and Faber, London (1955), pp. 61–62, 115.
12. C. DE JAGER, *Handbuch der Physik (Encyclopaedia of Physics)*, Springer, Berlin, Vol. 52 (1959), pp. 82–88, 98–99.
13. M. WALDMEIER, *Ergebnisse und Probleme der Sonnenforschung*, Akad. Verlagsges., Leipzig (1955), pp. 92–101.

14. M. Minnaert, *The Sun* (edited by G. P. Kuiper), Univ. of Chicago Press, Chicago, Ill. (1953), pp. 168–177.
15. K. O. Kiepenheuer, *The Sun* (edited by G. P. Kuiper), Univ. of Chicago Press, Chicago, Ill. (1953), pp. 365–376.
16. C. de Jager, *Handbuch der Physik (Encyclopaedia of Physics)*, Springer, Berlin, Vol. 52 (1959), pp. 173–182.
17. G. Abetti, *The Sun*, Faber and Faber, London (1955), pp. 67–68, 83–105.
18. G. Abetti, *The Sun*, Faber and Faber, London (1955), pp. 62–89.
19. C. de Jager, *Handbuch der Physik (Encyclopaedia of Physics)*, Springer, Berlin, Vol. 52 (1959), pp. 151–173.
20. M. Waldmeier, *Ergebnisse und Probleme der Sonnenforschung*, Akad. Verlagsges., Leipzig (1955), pp. 139–194.
21. M. A. Ellison, *The Sun and its Influence*, Routledge and Kegan Paul, London (1956), pp. 31–50.
22. K. O. Kiepenheuer, *The Sun* (edited by G. P. Kuiper), Univ. of Chicago Press, Chicago, Ill. (1953), pp. 323–361; M. Minnaert, *ibid.*, pp. 722–727.
23. H. C. van de Hulst, *The Sun* (edited by G. P. Kuiper), Univ. of Chicago Press, Chicago, Ill. (1953), p. 208.
24. G. Abetti, *The Sun*, Faber and Faber, London (1955), pp. 100–103.
25. G. Abetti, *The Sun*, Faber and Faber, London (1955), pp. 100–103, 119–121, 204–207, 241.
26. H. Friedman, Communication to the IAU at Berkeley, Calif. on August 16, 1961.
27. H. C. van de Hulst, *The Sun* (edited by G. P. Kuiper), Univ. of Chicago Press, Chicago, Ill. (1953), pp. 207–255; K. O. Kiepenheuer, *ibid.*, pp. 394–398; C. W. Allen, *ibid.*, pp. 648–649.
28. M. Waldmeier, *Ergebnisse und Probleme der Sonnenforschung*, Akad. Verlagsges., Leipzig (1955), pp. 211–251.
29. C. de Jager, *Handbuch der Physik (Encyclopaedia of Physics)*, Springer, Berlin, Vol. 52 (1959), pp. 80–135, 139–142, 149–155, 330–331.
30. *Ann. Astrophys. Obs. Smithsonian Inst.*, 6 (1942) 29–31, 33–42.
31. M. Nicolet, *Arch. Meteorol. Geophys. Bioklimatol. Ser. B*, 3 (1951) 209–219.
32. F. S. Johnson, *J. Meteorol.*, 11 (1954) 431–439.
33. R. Stair and R. G. Johnston, *J. Res. Natl. Bur. Std.*, 57 (1956) 205–211.
34. T. E. Stern and N. Dieter, *Smithsonian Contrib. Astrophys.*, 3, No. 3 (1958) 9–12.
35. C. G. Abbot, *Smithsonian Contrib. Astrophys.*, 3, No. 3 (1958) 13–21.
36. L. B. Aldrich and W. H. Hoover, *Science*, 116 (1952) 2.
37. F. E. Stuart, C. E. Sheldon Jr., E. P. Todd and P. R. Gast, *Geophys. Res. Papers (U.S.)*, October 1959.
38. *Ann. Astrophys. Obs. Smithsonian Inst.*, 5 (1932) 103–110.
39. *Ann. Astrophys. Obs. Smithsonian Inst.*, 6 (1942) 31, 35–36, 62.
40. *Ann. Astrophys. Obs. Smithsonian Inst.*, 5 (1932) 167.
41. *Ann. Astrophys. Obs. Smithsonian Inst.*, 6 (1942) 29–31, 33–37, 44, 63–64.
42. *Ann. Astrophys. Obs. Smithsonian Inst.*, 7 (1954) 18.
43. *Ann. Astrophys. Obs. Smithsonian Inst.*, 5 (1932) 110–119.
44. *Ann. Astrophys. Obs. Smithsonian Inst.*, 6 (1942) 9, 33, 64–65, 72.
45. *Ann. Astrophys. Obs. Smithsonian Inst.*, Vols. 5–7 (1932–1954).
46. H. L. Johnson and B. Iriarte, *Lowell Observatory IV*, No. 8, Bull. 96 (1959), pp. 99–104.
47. J. Georgi, *Z. Meteorol.*, 5 (1951) 136–139.
48. J. Georgi, *Meteorol. Rundschau*, 4 (1951) 193–194.
49. J. Georgi, *Meteorol. Rundschau*, 5 (1952) 65.

50. J. Georgi, *Meteorol. Rundschau*, 6 (1953) 64.
51. J. Georgi, *Meteorol. Rundschau*, 7 (1954) 182–184.
52. J. Georgi, *Ann. Meteorol.*, 5 (1952) 83–96.
53. W. Schüepp, *Geofis. Pura Appl.*, 30 (1955) 179–181.
54. K. Wegener, *Geofis. Pura Appl.*, 15 (1949) 1–4.
55. K. Wegener, *Geofis. Pura Appl.*, 16 (1950) 1–4.
56. K. Wegener, *Meteoros (Buenos Aires)*, 1 (1951) 171–182.
57. M. Waldmeier, *Sonne und Erde*, 2. Auflage, Büchergilde Gutenberg, Zürich (1946).

Chapter 2

THE ASTRONOMICAL AND GEOGRAPHICAL FACTORS AFFECTING THE AMOUNT OF SOLAR RADIATION REACHING THE EARTH*

1. THE SUN–EARTH DISTANCE

The earth describes an ellipse round the sun, with the latter at one of the foci. The apparent path of the sun as seen from the earth is known as the ecliptic.

The eccentricity of the earth's orbit is very small ($e=0.01673$), so that the orbit is in fact very nearly circular. Thus, the shortest distance (when the earth is in perihelion) and the longest distance (when the earth is in aphelion) are respectively given by

$$R_p = a(1-e) = 147.10 \times 10^6 \text{ km}$$

and   (2.1)

$$R_a = a(1+e) = 152.10 \times 10^6 \text{ km}$$

where a is the semi-major axis of the earth's orbit. The so-called mean distance of the earth from the sun is defined as half the sum of perihelion and aphelion distances and is therefore equal to a. The numerical value of this quantity is used as a unit of distance and is called the 'astronomical unit' (A.U.). The astronomical unit may be expressed in kilometres, or in angular measure, in terms of the solar parallax $\pi_\odot$, which is the angle subtended at the centre of the sun by the equatorial radius of the earth.

The solar parallax can be determined by various methods; some of the results are given in Table 2.1. The internationally accepted value is $\pi_\odot = 8.80''$ $= 4.263 \times 10^{-5}$ radians, which gives 1 A.U. $= 149.5 \times 10^6$ km. An error of $0.01''$ in the solar parallax causes an error of 170,000 km in the A.U.

It follows from the above that the radius vector ($R.V.$) of the sun, its apparent radius $R_\odot$, and the intensity of solar radiation at the earth's

---

* The influence of the atmosphere will be disregarded here (see Chapter 3).

TABLE 2.1. Values of the solar parallax $\pi_\odot$ obtained by different methods.

Method	$\pi_\odot$	A.U. (km)
Radar	—	149,498,000*
Eros opposition	8.790″ ± 0.001″	149,670,000
Earth mass	8.798″ ± 0.0004″	149,534,000
Variation of $\pi_\odot$	8.793″ ± 0.003″	149,619,000
Aberration	8.79″	149,670,000
Accepted value	8.80″	149,500,000

* Accuracy not known.

TABLE 2.2. Relative intensity of radiation on the first of each month, without allowance for the effect of the atmosphere.

Date	I_t/I_0
1 January	1.0335
1 February	1.0288
1 March	1.0173
1 April	1.0009
1 May	0.9841
1 June	0.9714
1 July	0.9666
1 August	0.9709
1 September	0.9828
1 October	0.9995
1 November	1.0164
1 December	1.0288

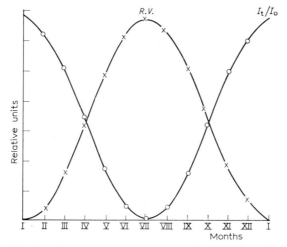

Fig. 2.1. The relative intensity of radiation I_t/I_0 and the radius vector of the sun $R.V.$ as a function of time.

TABLE 2.3. Variation in the radius vector of the sun and in the apparent radius of the sun (1961).

Date		Radius vector	Radius	Date		Radius vector	Radius
Jan.	1	0.9832	16′17.59″	July	1	1.0167	15′45.42″
	10	0.9834	17.43		10	1.0167	45.42
	20	0.9840	16.80		20	1.0161	45.90
Feb.	1	0.9853	15.48	Aug.	1	1.0149	47.05
	10	0.9868	14.00		10	1.0137	48.23
	20	0.9888	12.02		20	1.0118	49.98
Mar.	1	0.9909	10.03	Sept.	1	1.0091	52.48
	10	0.9932	7.75		10	1.0069	54.57
	20	0.9960	5.06		20	1.0042	57.17
Apr.	1	0.9993	1.81	Oct.	1	1.0011	16′ 0.13″
	10	1.0020	15′59.28″		10	0.9985	2.58
	20	1.0048	56.60		20	0.9957	5.37
May	1	1.0077	53.80	Nov.	1	0.9924	8.49
	10	1.0099	51.78		10	0.9902	10.65
	20	1.0120	49.76		20	0.9880	12.86
June	1	1.0141	47.83	Dec.	1	0.9860	14.81
	10	1.0153	46.66		10	0.9848	16.04
	20	1.0162	45.83		20	0.9838	17.04

surface, must all depend on the earth's position in its orbit. Table 2.2 and Fig. 2.1 show the ratio of the radiation intensity I_t on the first of every month and the corresponding intensity I_0 at the mean sun–earth distance.

Table 2.3 shows the variation in $R.V.$ and $R_\odot$ for 1961 in angular units. The maximum value of $R.V.$ is found to occur at the beginning of July (cf. Table 2.2 and Fig. 2.1) and the minimum at the beginning of January. The exact values can be found in an ephemeris [1].

2. COORDINATE SYSTEMS OF THE SUN AS A CELESTIAL BODY

The position of the sun on the celestial sphere can be specified by means of a number of coordinate systems. The horizontal and the equatorial systems are commonly employed.

In the horizontal system the position of the sun is specified by the zenith distance (or its complement, the altitude of the sun) and the azimuth. These coordinates are defined as follows. Consider a line drawn through the observer in the direction of the plumb-line. The line cuts the celestial sphere at two points: the zenith vertically above the observer (Z in Fig. 2.2) and the nadir vertically below him. The plane drawn through the same observer at right angles to the zenith–nadir line cuts the celestial sphere in the apparent horizon. The plane parallel to this horizon, and passing through the earth's centre, is the true horizon (H—H in Fig. 2.2). Small circles parallel to the horizon are called azimuth circles.

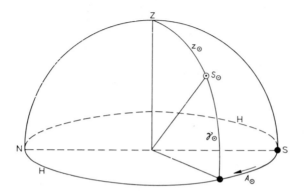

Fig. 2.2. The horizontal coordinate system.

The zenith distance $z_\odot$ of the sun is its angular distance from the zenith in the vertical circle containing the zenith, the nadir and the sun. The altitude of the sun is then $\gamma_\odot = 90° - z_\odot$. The azimuth $A_\odot$ is the angular distance between the vertical circle containing the zenith and the sun, and the south point (S in Fig. 2.2). It is measured from the south through west, north and east. S is the observer's 'geographical south'.

The coordinates of the sun in the horizontal system, which may be referred to as a 'subjective system', depend on the position of the observer and on the time. It is possible to construct a solar chart giving values of $\gamma_\odot$ and $A_\odot$ for a given geographical latitude φ.

In the equatorial system, which may be referred to as an 'objective system', the position of the sun is specified by the polar distance $P_\odot$ (or its complement, the declination $\delta_\odot$) and the right ascension, $R.A._\odot$. In this system the basic points are the celestial poles P_N and P_S at which the earth's axis intersects the celestial sphere (Fig. 2.3). The basic circles are the celestial equator and the hour circle. The former is cut on the celestial sphere by the plane passing through the earth's centre and perpendicular to its axis, and the latter is the great circle on the celestial sphere which passes through P_N, P_S and the sun, cutting the equator at 90°. The polar distance $P_\odot$ is the

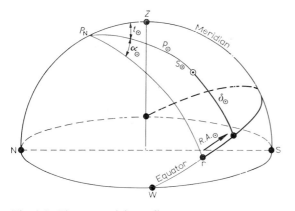

Fig. 2.3. The equatorial coordinate system.

angular distance from P_N, measured along the hour circle of the sun. It is clear that $\delta_\odot = 90° - P_\odot$. The right ascension, $R.A._\odot$, is the angular distance of the hour circle of the sun from the vernal equinox (Υ in Fig. 2.3), measured in the westward direction in degrees or hours. The vernal equinox is the point of intersection of the ecliptic and the celestial equator, and is passed by the sun in the spring. There is also the analogous autumnal equinox.

The angle $t_\odot$ in Fig. 2.3 can also be used as a coordinate in the equatorial system. It is defined as the angle between the great circle through P_N, Z and P_S, (*i.e.* the meridian) and the hour circle, measured in the direction of the daily movement in degrees or hours.

The point Υ at which the sun is seen from the earth at the vernal equinox is also called the first point of Aries, but is not marked on the sky

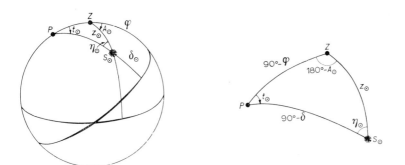

Fig. 2.4. Illustrating the transformation of coordinates.

by a celestial body. Evidently, Υ can also be defined by an hour angle $\theta_\odot = t_\odot + \alpha_\odot$. When the sun is in the meridian, at true noon, $A_\odot = 0$ (Fig. 2.2), $t_\odot = 0$ (Fig. 2.3) and therefore $\theta_\odot = \alpha_\odot$. This defines the 'sidereal time' (see p. 44).

The conversion from one of the coordinate systems to the other may be achieved with the aid of spherical trigonometry. Thus, consider the triangle whose apices at the pole, the zenith and the sun are connected by great circles as shown in Fig. 2.4. In the conversion from the first system into the second φ, $z_\odot$ and $A_\odot$ are known, and $\delta_\odot$, $t_\odot$ and $\eta_\odot$ are computed using the formulae*:

a) $\cos \delta_\odot \sin t_\odot = \sin z_\odot \sin A_\odot$

b) $\cos \delta_\odot \cos t_\odot = \cos \varphi \; \cos z_\odot + \sin \varphi \sin z_\odot \cos A_\odot$

c) $\qquad \sin \delta_\odot = \sin \varphi \; \cos z_\odot - \cos \varphi \sin z_\odot \cos A_\odot$ $\qquad$ (2.2)

d) $\cos \delta_\odot \sin \eta_\odot = \cos \varphi \; \sin A_\odot$

e) $\cos \delta_\odot \cos \eta_\odot = \sin \varphi \; \sin z_\odot + \cos \varphi \cos z_\odot \cos A_\odot$

In the conversion from the second system into the first, φ, $\delta_\odot$ and $t_\odot$ are given and $z_\odot$, $A_\odot$ and $\eta_\odot$ are computed with the aid of:

a) $\sin z_\odot \sin A_\odot = \cos \delta_\odot \sin t_\odot$

b) $\qquad \cos z_\odot = \sin \varphi \sin \delta_\odot + \cos \varphi \cos \delta_\odot \cos t_\odot$

c) $\sin z_\odot \cos A_\odot = -\cos \varphi \sin \delta_\odot + \sin \varphi \cos \delta_\odot \cos t_\odot$ $\qquad$ (2.3)

d) $\sin z_\odot \sin \eta_\odot = \cos \varphi \sin t_\odot$

e) $\sin z_\odot \cos \eta_\odot = \sin \varphi \cos \delta_\odot - \cos \varphi \sin \delta_\odot \cos t_\odot$

3. COMPUTATION OF INSOLATION ON DIFFERENT PLANES AT THE EARTH'S SURFACE

The quantity of energy insolated on to any given plane at the earth's surface can be computed with the aid of Eqs. (2.2) and (2.3) by integrating within the appropriate time limits. However, such trigonometrical computations are of minor practical value since they neglect atmospheric effects. The atmosphere scatters a part of the radiation emitted by the sun and, depending on the astronomical and atmospheric conditions, gives rise to attenuation of the direct solar radiation. These effects and the results of calculations of real radiation intensities are further discussed in Chapters 3 and 4.

---

* All formulae given in this chapter are reproduced without derivations since these may be found in books on spherical trigonometry or astronomy.

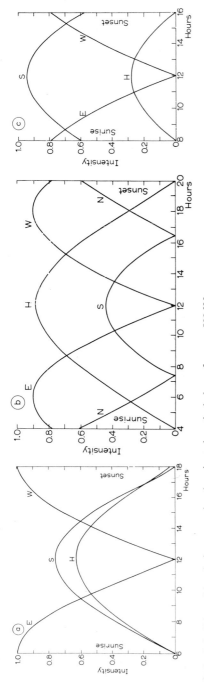

Fig. 2.5. Diurnal insolation on variously oriented vertical planes for $\varphi = 50°$ [2].
a = equinox, b = summer solstice, c = winter solstice.

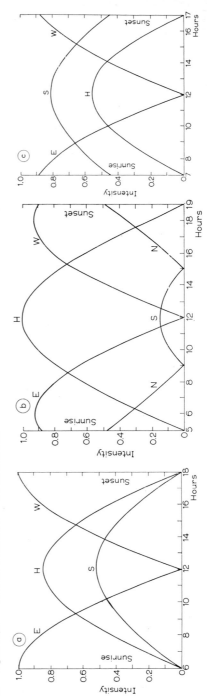

Fig. 2.6. Diurnal insolation on variously oriented vertical planes for $\varphi = 32°$ [2].
a = equinox, b = summer solstice, c = winter solstice.

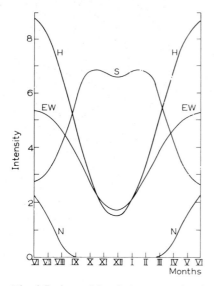

Fig. 2.7. Annual insolation on variously oriented vertical planes for $\varphi = 50°$ [2].

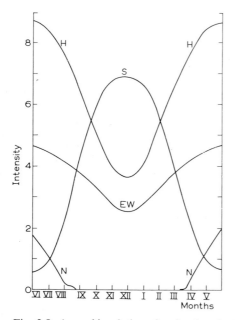

Fig. 2.8. Annual insolation of variously oriented vertical planes for $\varphi = 32°$ [2].

In order to evaluate the role of astronomical and geographical factors we shall consider those calculations and results which exhibit the general features of the variation in the insolation.

The insolation on a plane normal to the solar beam is a function of the sun–earth distance only, as shown in Fig. 2.1 and Table 2.1. In our further discussion we shall use the solar constant I_0 so that variations with distance will be excluded. The intensity I_H on a horizontal plane is given by

$$I_H = I_0 \cos z_\odot = I_0 \sin \gamma_\odot \qquad (2.4)$$

where $\gamma_\odot$ is the altitude of the sun and is equal to the angle between the solar beam and the insolated plane. For a fixed point of observation φ is constant, and for a given day it may be assumed that $\delta_\odot$ is also constant. The diurnal variation of I_H for December 21, March 21, and June 21 at $\varphi = 50°$ (Central Europe, U.S.A., Canada) and $\varphi = 32°$ (Israel) is given in Figs. 2.5 and 2.6.

The daily quantities of insolation energy may be calculated by integrating the curves of Figs. 2.5 and 2.6; the results are shown in Figs. 2.7 and 2.8. The quantity insolated per m² per hour at normal incidence is taken as the unit.

For vertical planes facing east and west, the intensities I_E and I_W are zero at real noon when the sun is in the south (curves marked EW).

The curves marked N and S which represent the variation of I_N and I_S are of particular interest. As can be seen, although in the summer $I_N(50°) > I_N(32°)$, the period of insolation for $\varphi = 50°$ is less than that for $\varphi = 32°$,

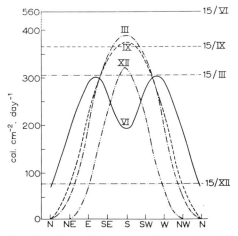

Fig. 2.9. Diurnal insolation of variously oriented vertical planes and on the horizontal plane on solstices and equinoxes [3].

TABLE 2.4. Quantity of insolated energy per day on a horizontal plane, without allowance for the influence of the atmosphere [3].

Date	North										South								
Latitude (degrees)	90	80	70	60	50	40	30	20	10	0	10	20	30	40	50	60	70	80	90
Mar. 21		160	316	461	593	707	799	867	909	923	909	867	799	707	593	461	316	160	0
Apr. 13	436	436	541	655	755	832	892	922	925	900	849	773	674	555	421	277	131	7	0
May 6	796	784	772	834	894	938	958	952	921	863	783	680	560	426	285	144	24	0	0
May 29	1030	1014	968	963	988	1002	997	964	908	829	729	611	479	339	199	70	0	0	0
June 22	1110	1093	1043	1009	1020	1022	1005	964	900	814	708	585	450	306	170	48	0	0	0
July 15	1025	1010	963	958	983	1007	990	959	904	825	726	608	477	338	198	70	0	0	0
Aug. 8	789	777	765	826	886	929	949	944	913	856	776	674	555	422	282	143	24	0	0
Aug. 31	431	431	535	648	747	823	882	911	914	890	839	764	666	549	417	274	130	7	
Sept. 23		158	312	456	586	698	789	857	898	912	898	857	789	698	586	456	312	158	
Oct. 16		7	133	281	427	562	683	783	861	913	938	935	904	844	766	664	548	442	442
Nov. 8			25	150	295	442	581	706	813	897	956	989	994	973	929	866	802	814	826
Nov. 30				74	210	359	507	646	771	877	960	1019	1052	1059	1045	1018	1024	1073	1089
Dec. 22				51	181	327	480	624	756	869	962	1030	1073	1092	1089	1078	1114	1167	1185
Jan. 13				75	211	361	509	649	775	881	965	1024	1057	1064	1050	1023	1029	1078	1095
Feb. 4			25	151	298	447	586	712	820	905	965	998	1003	982	937	873	809	821	834
Feb. 26		7	135	285	432	570	691	793	871	924	949	946	915	854	775	672	556	447	447

i.e. seven hours per day instead of eight. I_s has a minimum in the summer and a maximum in the winter.

Fig. 2.9 shows the daily quantities of energy insolated on vertical planes passing N, NE, E, SE, S, SW, W and NW and on the horizontal plane for December 15, March 15, June 15 and September 15 (in cal · cm^{-2} · day^{-1}).

Table 2.4 gives the quantity of energy insolated per day on a horizontal plane for different φ and dates. These data are also plotted in a three-dimensional form in Fig. 2.10. The value of φ is plotted along the x-axis, the date is plotted along the y-axis, and the amount of energy along the z-axis.

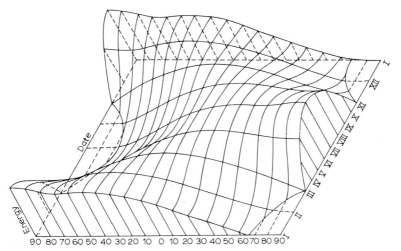

Fig. 2.10. A three-dimensional diagram of diurnal insolation [3].

Fig. 2.11 shows the insolation intensities on the 15th of every month at Jerusalem on vertical planes facing N, NE, E, SE, S, SW, W, and NW, on the horizontal plane H and on a plane normal to the solar beam.

A method of computation and demonstration of direct solar radiation falling on a building as a whole has been given elsewhere [4,5].

So far, only the horizontal and vertical planes have been discussed. Planes inclined at various angles to the horizon are of equal importance. In general, the intensity I_p incident on a given plane is given by

$$I_p = I_0 \cos \beta \tag{2.5}$$

where β is the angle between the solar beam and the normal to the plane under consideration. This angle is of course a function of $z_\odot$, A and the angle

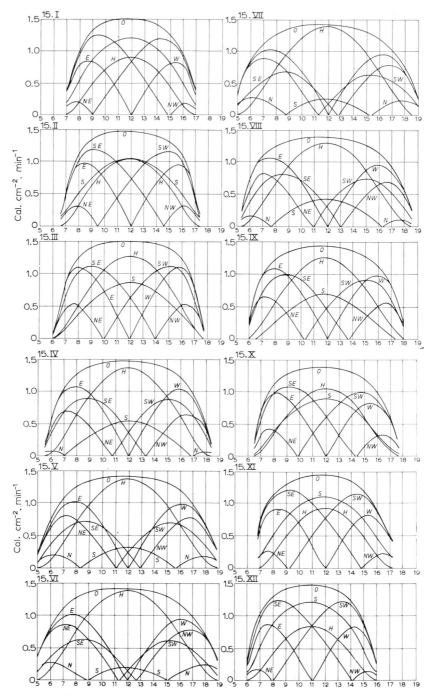

Fig. 2.11. Diurnal variation in insolated energy on various planes on the 15th of every month at Jerusalem (Courtesy of Prof. Sh. Irmay, Technion, Haifa).

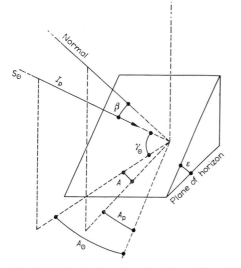

Fig. 2.12. Illustrating the calculation of insolation on an arbitrarily inclined plane.

ε between the given plane and the horizon. This is made clear in Fig. 2.12, from which it is evident that A is the difference between the azimuth of the sun $A_\odot$ and the azimuth A_p of the vertical plane containing the normal to the given inclined plane.

The relation between the various angles is

$$\cos \beta = \cos \varepsilon \sin \gamma_\odot + \sin \varepsilon \cos \gamma_\odot \cos A \qquad (2.6)$$

and hence, using Eqs. (2.3b), (2.2c) and (2.2a) we have:

$$\begin{aligned}
I_p = I_0 [&\cos \varepsilon (\sin \varphi \sin \delta_\odot + \cos \varphi \cos \delta_\odot \cos t_\odot) \\
&+ \sin \varepsilon \{\cos A_p [\tan \varphi (\sin \varphi \sin \delta_\odot + \cos \varphi \cos \delta_\odot \cos t_\odot) \qquad (2.7) \\
&- \sin \delta_\odot \sec \varphi] + \sin A_p \cos \delta_\odot \sin t_\odot \}]
\end{aligned}$$

The intensities on horizontal and vertical planes can be obtained by setting $\varepsilon = 0$ and $\varepsilon = 90°$ respectively.

4. HELIOGRAPHIC COORDINATES

The various coordinates discussed above are used in defining the position of the sun on the celestial sphere and for other astronomical purposes. In order to follow a phenomenon on the surface of the sun it is necessary to define heliographic coordinates which specify the position of any point on the surface of the sun. This is done by analogy with the definition of the terrestrial latitude and longitude.

The fundamental points in the heliographic system are the poles, *i.e.* the intersections of the axis of rotation of the sun with its surface (Fig. 2.13). The sun's axis exhibits almost no precession. The plane perpendicular to this axis and passing through the centre of the sun cuts a great circle on the sun's surface. This circle is the solar equator and serves as a reference for the determination of the heliographic latitude $\varphi_\odot$.

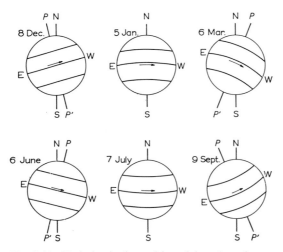

Fig. 2.13. Variation in the position of the solar axis.

In order to define the heliographic longitude $L_\odot$ an arbitrary zero meridian must be chosen. This has been defined as the meridian passing through the point of intersection of the solar equator with the ecliptic on January 1, 1854 at 12 h U.T. (see p. 45). The longitude can assume values between $0°$ and $360°$ and is measured in the direction of the sun's rotation.

A mean sidereal period of rotation of the sun of 25.38 days is adopted as the rotation period for $\varphi_\odot = 16°$. A mean period is used because the sun does not revolve as a solid body, the rate of rotation decreasing with increasing $\varphi_\odot$. By adopting this rotation period, taking the 'zero' rotation number as that on January 1, 1854 at 12 h U.T. (the rotation number increases by one every 25.38 days), and assuming the mean rotation rate to be 14.1844° per day, the longitude $L_\odot$ of a certain point on the sun can be defined.

In general, the dependence of the angular sidereal motion of the sun on $\varphi_\odot$ is

$$14.38° - 2.77° \sin^2 \varphi_\odot \quad \text{(per day)} \tag{2.8}$$

This expression is somewhat different for different solar phenomena.

This method of rotation counting is known as the Carrington counting and is used in astronomical yearbooks. The value of the heliographic longitude $L_\odot$ for the centre of the visible solar disc is available in tabulated form as a function of time.

The synodic rotation period is found to be convenient in solar observations. A mean value of 27.2753 days based on the mean sidereal value of 25.38 is accepted. The dependence of the angular motion on $\varphi_\odot$ is then

$$13.29° - 2.7° \sin^2 \varphi_\odot \quad \text{(per day)} \tag{2.9}$$

$$26.90 + 5.2 \sin^2 \varphi_\odot \quad \text{(days)} \tag{2.10}$$

However, it is more convenient in practice to use a synodic period of an integral number of days, namely 27, corresponding to $\varphi_\odot = 8°$. On this method of counting $L_\odot$ was zero on February 8, 1832 at 0 h U.T.

Although the solar axis is fixed in space, to a terrestrial observer it appears to change its position in relation to the celestial axis. This apparent motion is due to the finite angle i between the solar meridian and the ecliptic ($i = 7°15'$; see Fig. 2.14). In exact calculations the following expression must be employed

$$i = 7.252° + 0.00002°(t - 1950) \tag{2.11}$$

where t is the year of observation. The longitude of the point of intersection Ω of the solar equator and the ecliptic (Fig. 2.14) is given by

$$L_{\odot\Omega} = 74.962° + 0.01295°(t - 1950) \tag{2.12}$$

In the ephemeris $i = 7.25°$ and $L_{\odot\Omega} = 75.063° + 0.01396°(t - 1950)$ are employed [1].

A terrestrial observer sees both solar poles just on the edge of the solar disc only at the beginning of June and the beginning of December (Fig. 2.13). At all other times only the north or the south pole is visible.

It is evident from Fig. 2.13 that the heliographic latitude $\varphi_\odot$ of the

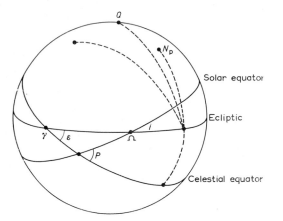

Fig. 2.14. The mutual disposition of the solar equator, the celestial equator and the ecliptic.

centre of the solar disc is not constant. Tabulated values of this latitude are available in astronomical yearbooks.

Finally, a further quantity which is used is the position angle of the sun which is defined as the angle between the celestial and the solar equators (Fig. 2.14). This is taken as positive in the north–east direction.

5. TIME AND ITS UNITS

Years and days are employed as units for the measurement of time. Five different years are used. These are the tropical year (the time between two similar equinoxes), the sidereal year (the interval between two successive transits of the sun through the same point relative to the fixed stars), the anomalistic year (the interval between two successive transits of the sun through the perihelion), the Julian calendar year (based on a tropical year of 365.25 days) and the Gregorian calendar year (the civil calendar year). In this text we shall use the tropical year.

The 'day' is defined as the period between successive passages of the sun through a given meridian, and is referred to as the real solar day. For a number of reasons, two of which will be mentioned here, it is not constant.

Because of the eccentricity of its orbit, the angular velocity of the earth round the sun is not constant. It follows from Kepler's laws that the maximum velocity occurs at the perihelion and the minimum at the aphelion.

It is therefore evident that the maximum time period between two consecutive transits of the sun across a given meridian occurs near the aphelion. The difference between these periods and the mean solar day is the so-called equation of time due to eccentricity (see curve 1 in Fig. 2.15).

The second reason for the variation in the length of the real solar day is due to the fact that the ecliptic is inclined to the celestial equator. The angle between the ecliptic and the equator (ε in Fig. 2.14) is approximately $23°27'8.2''$. The angular velocity of the earth relative to the hour circles (the meridians) between the spring equinox and the summer solstice is less than the angular velocity between the summer solstice and the spring equinox, and so on.

The difference between the real and the mean solar day, which is due to the obliquity, is the so-called equation of time due to obliquity (curve 2 in Fig. 2.15).

The two partial equations referred to above define the overall equation of time (curve 3 in Fig. 2.15). Finally, the angle ε is also a function of time.

For some time the angular velocity of the earth around its axis was assumed to be constant and this was used to define the mean solar day by dividing the year into equal parts. This unit of time was based on the times of transit of the sun through the Greenwich meridian and was referred to as 'Greenwich Mean Time' (G.M.T.) and later 'Universal Time' (U.T.). Because of deceleration and unevenness in the earth's rotation, the day based on the Universal Time is not constant.

Recently, a further (constant) mean day was introduced and named the 'Ephemeris Day' (it was referred to as the Newtonian Day until 1952). The Ephemeris Day can be realized with the aid of a highly accurate chro-

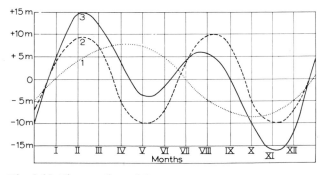

Fig. 2.15. The equation of time.

References p. 46

nometer, but Ephemeris Time does not define the actual position of the sun on the celestial sphere for a terrestrial observer.

Clocks in daily use are regulated in accordance with the Zone Mean Time which is taken as constant in 15° intervals of geographic longitude, starting from Greenwich. This has to be taken into consideration in converting from Local Mean Time into Zone Mean Time and vice versa. Before the equation of time is used, Z.M.T. has to be converted into L.M.T. For example, Haifa has a longitude of 35° and the Zone Mean Time is that of Alexandria ($\varphi = 30°$). The difference of 5° is equivalent to 20 minutes, *i.e.* 11 h 40 m Z.M.T. is the real noon for Haifa. The equation of time must then be used in order to calculate the real solar time.

REFERENCES

1. *The Astronomical Ephemeris*, H.M. Stationery Office, London (1961).
2. A. PALOTAI AND J. BRÜNN, *Gesundh. Ing.*, 54 (1931) 122.
3. K. YA. KONDRATYEV, *Radiant Solar Energy* (in Russian), Leningrad (1954), pp. 264, 330, 592.
4. N. ROBINSON AND M. PELEG, *Biometeorology (Proc. 2nd Congr. Intern. Soc. Biometeorol., 1960)* (edited by S. W. Tromp), Pergamon, London (1962), pp. 262–269.
5. M. PELEG, *Bull. Sta. Tech. Climatol. Haifa*, No. 5 (1956).

Chapter 3

THE EFFECT OF THE ATMOSPHERE ON SOLAR RADIATION REACHING THE EARTH

1. INTRODUCTION

The intensity of solar radiation outside the atmosphere was discussed in the preceding chapter. It was shown that this radiation is subject to seasonal and diurnal variations which are due to astronomical effects. Moreover, the amount of radiation emitted by the sun is a function of wavelength and depends on the presence of solar flares, sunspots, and so on. Effects associated with the terrestrial atmosphere are superimposed on these variations. Thus, the energy reaching the earth's surface is substantially modified by scattering and absorption in the atmosphere.

For general and more extensive information on the effect of the atmosphere, see [102–106].

The depletion of the incident solar energy may be characterized by the extinction coefficient a_λ which is a function of the wavelength λ and represents both scattering and absorption. Let us discuss the scattering first.

Consider a parallel beam of light incident on a single particle. If only scattering occurs then all the energy in the original beam will be present in the radiation field surrounding the particle, but it will be dispersed, so that some of it will travel in all directions. In effect, the particle thus becomes a new source of light. For large particles this change in direction may be due to a number of causes, for example diffraction, reflection, refraction, or a combination of these effects.

Absorption is said to occur if the sum of the scattered and transmitted energies integrated over all angles θ is less than the incident energy. The absorbed energy may alter the temperature, the chemical composition, and a number of other properties of the particle.

The change dI_λ in the intensity I_λ of a parallel monochromatic beam is given by Beer's law

$$dI_\lambda = -I_\lambda a'_\lambda dx = -I_\lambda a''_\lambda \varrho dx \tag{3.1}$$

where ϱ is the density of the medium, dx is the path length, a'_λ is the extinction coefficient per unit length, and a''_λ is the extinction coefficient per unit mass. Thus in a homogeneous medium equal increments of length dx will deplete equal fractions dI_λ/I_λ of the incident energy. Integration of Eq. (3.1) yields

$$I_\lambda = I_{0\lambda} \exp\left[-\int_0^x a'_\lambda dx\right] = I_{0\lambda} \exp\left[-\int_0^x a''_\lambda \varrho\,dx\right] \tag{3.2}$$

The absorbing and scattering material in the atmosphere varies along the path of the solar beam because the gaseous composition of the atmosphere is a function of height and time. This also applies to particulate matter, including clouds. Finally, each atmospheric constituent has its own scattering and absorption coefficients which also vary with the wavelength.

2. PATH LENGTH IN THE ATMOSPHERE

When the sun is not at the zenith, the intensity of solar radiation reaching the earth's surface is given by (cf. Eq. (3.2))

$$I_\lambda = I_{0\lambda} \exp\left[-\int_0^S a''_\lambda \varrho\,dS\right] \tag{3.3}$$

where the integration is carried out along the path S of the radiation. In general, S will be a curve. Since the relation between the path elements along the vertical and along S is

$$dS = -m_r dh \tag{3.4}$$

where the negative sign implies that S and h are measured in opposite directions, and since for air $d\varrho = -\varrho g\,dh$, we have on substituting into Eq. (3.3)

$$I_\lambda = I_{0\lambda} \exp\left[\int_0^\varrho (a''_\lambda \cdot m_r/g)d\varrho\right] \tag{3.5}$$

The quantity m_r which enters into Eq. (3.5) is the so-called optical air mass. It is clear that in a plane-stratified atmosphere in which refraction may be neglected

$$dS = -\sec z_\odot\,dh$$

where $z_\odot$ is the zenith angle (see Fig. 3.1), and hence

$$m_r = \sec z_\odot \tag{3.6}$$

This shows that m_r is in fact the relative air mass, because on this approximation we also have

$$-\frac{\int a''_\lambda \varrho \, dS}{\int a''_\lambda \varrho \, dh} = \frac{\int a''_\lambda \varrho \sec z_\odot dh}{\int a''_\lambda \varrho \, dh} = \sec z_\odot \tag{3.7}$$

i.e. when refraction may be neglected, $\sec z_\odot$ is equal to the ratio of the optical path lengths in the oblique and vertical directions. When the extinction coefficient a''_λ may be regarded as constant for the given λ this ratio becomes equal to the ratio of the actual masses of air contained in cylinders of unit cross-section and lying along the two directions.

We are thus led to a more general definition of m_r, namely

$$m_r = \frac{\int a''_\lambda \varrho \, dS}{\int a''_\lambda \varrho \, dh} \tag{3.8}$$

which reduces to

$$m_r = \frac{\int \varrho \, dS}{\int \varrho \, dh} \tag{3.9}$$

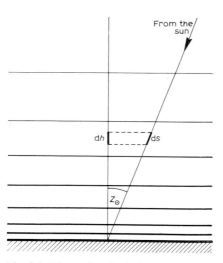

Fig. 3.1. Illustrating the derivation of Eq. (3.6).

when a_λ'' is a function of λ only. Eq. (3.3) may now be rewritten in the form

$$I_\lambda = I_{0\lambda} \exp(-a_\lambda m_r) \tag{3.10}$$

where $a_\lambda = \int a_\lambda'' \varrho \, dh$ and is the total extinction coefficient in the vertical direction, while m_r is the relative air mass in the direction under investigation.

Another commonly employed form of Eq. (3.10) is

$$I_\lambda/I_{0\lambda} = (q_\lambda)^{m_r} \tag{3.11}$$

where q_λ is the transmissivity or the transmission factor of the atmosphere.

It is useful to separate a_λ'' into two components such that

$$a_\lambda'' = \sigma_\lambda'' + \alpha_\lambda''$$

where σ_λ'' is the scattering coefficient and α_λ'' is the absorption coefficient at the given λ. It is found that they vary with the type of the absorbing or scattering material.

Finally, the depletion of forward moving radiation will also occur by reflection at air–solid and air–liquid interfaces; this is particularly important at air–ocean and air–ground boundaries, especially when the surface is covered by ice or snow.

A further useful idea is that of the homogeneous atmosphere with a height $H = p_0/g\varrho_0$, where p_0 and ϱ_0 are the pressure and density at the earth's surface. On substituting $p_0 = 760$ mm Hg and ϱ_0 corresponding to the temperature of 273°K, it is found that $H = 7991$ km. At $T = 15°C$ a similar calculation yields $H = 8430$ km.

The air mass of a homogeneous atmosphere can easily be calculated, even with allowance for the curvature of the earth. Thus, the air mass of a

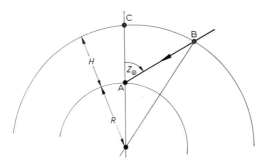

Fig. 3.2. Schematic representation of a solar beam traversing the atmosphere and directed at the earth's surface.

homogeneous spherical atmosphere (Fig. 3.2) is given by

$$m_{gr} = (BA)/(CA) = [\{(R/H)\cos z_\odot\}^2 + 2(R/H) + 1]^{\frac{1}{2}} - (R/H)\cos z_\odot$$
(3.12)

where R is the radius of the earth. The values of m_{gr} for $H = 7991$ km and $R = 6370$ km are listed in Table 3.1, showing that m_{gr} and $\sec z_\odot$ are very nearly equal (to two decimal places) for $z_\odot < 70°$.

TABLE 3.1. Values of optical air mass for various assumptions.

$z_\odot$	$\sec z_\odot$	m_{gr}	m_r (Bemporad)	$\sec z_\odot$ for $h = 22$ km
0	1.00	1.00	1.00	1.00
30	1.15	1.15	1.15	1.15
60	2.00	2.00	2.00	1.98
70	2.92	2.92	2.90	2.85
80	5.76	5.63	5.60	5.21
85	11.47	10.69	10.39	8.82
86	14.34	12.87	12.44	—
87	19.10	16.04	15.36	—
88	28.65	20.87	19.79	—
89	57.30	28.35	26.96	—
90	∞	39.94	39.70	12.06

In the actual atmosphere refraction and the decrease of density with height cannot be neglected. The application of the law of refraction as used in astronomy yields the following result [1]:

$$m_r = (\varrho_0 H)^{-1} \int_0^\infty \left[1 - \left(\frac{R}{R+h} \cdot \frac{n_0}{n} \right)^2 \sin^2 z_\odot \right]^{-\frac{1}{2}} \varrho \, dh$$
(3.13)

where n is the refractive index and $(n_0 - 1) \simeq 0.00029$ but is a function of the wavelength. The refractive index thus varies with ϱ and λ. Since both n and ϱ also vary with height h, and the variation depends on the atmospheric temperature and pressure, the integral (3.13) cannot be evaluated exactly.

Laplace assumed an isothermal atmosphere and found that

$$m_r = K(Refr/\sin z_\odot)$$

where $K = 1/c\varrho_0 n_0$, $Refr$ is the astronomical refraction, and c is the specific refraction in Newton's formula $n^2 - 1 = 2c\varrho$ which relates the refractive index and the density ϱ. The astronomical refraction is the number of degrees through which a ray is bent from its original rectilinear path on passing through the atmosphere. However, the assumption that the atmosphere is isothermal has had to be modified, and the most frequently used values of the air mass are those due to Bemporad who based his computations on early balloon soundings. Bemporad's values are also given in Table 3.1 from which it is clear that m_{gr} is a good approximation to Bemporad's results for $0 < z_\odot < 80°$. The agreement is still reasonable even for $z_\odot > 80°$ (5% at $z_\odot = 89°$).

A further complication is due to the fact that at elevated, e.g. mountain stations, the mass of the atmosphere above the observer is reduced. Even at sea level the pressure, and therefore the air mass, may vary by up to 10% from the average value, which must be taken into account in precise work. The absolute actual air mass may be estimated from

$$m/p = m_r/p_0 \qquad\qquad (3.14)$$

where p is the station pressure and p_0 the normal sea-level pressure.

This relation holds for a molecular atmosphere in which the scattering material is uniformly distributed. However, if the scattering material or absorbing matter is localized in a restricted range of elevations, the air mass must be corrected accordingly. This applies to water vapour which is located mainly in the lower atmosphere, to ozone which is located mainly in the upper atmosphere, to dust which may be distributed in layers, and so on. On the other hand Eq. (3.12) shows that for large $z_\odot$, $m_{gr} \simeq (1 + 2R/H)^{\frac{1}{2}}$, and hence m_{gr} increases with decreasing H. Thus, for water vapour and other depleting materials which are located near the bottom of the atmosphere, the optical air mass approaches the $\sec z_\odot$ law.

For a gas such as ozone, which can be approximately represented by a layer of depleting material at a high elevation, the optical air mass may be estimated as follows. Assuming that the zenith distance of the sun at height h is z_h, then

$$\frac{\sin z_\odot}{R+h} = \frac{\sin z_h}{R}$$

and hence

$$\sec z_h = \frac{1+h/R}{[\cos^2 z_\odot + 2(h/R)]^{\frac{1}{2}}} \simeq m_h \qquad\qquad (3.15)$$

The results of calculations based on this expression are given in Table 3.1 for $h = 22$ km.

3. SCATTERING

The depletion of energy in the direct solar beam is partly due to scattering by air molecules. If $I_{0\lambda}$ represents the intensity incident on a unit area of the surface of a plane homogeneous atmosphere, then the energy transmitted in the direct beam is given by

$$I_\lambda = I_{0\lambda} \exp(-\sigma'_\lambda m_r H) \qquad (3.16)$$

where σ'_λ is the scattering coefficient per unit length.

When the scattering particles are spherical, and if the radius r is small compared with λ, i.e. $r < 0.1\lambda$, and if it may be assumed that they scatter independently of one another, then the scattering coefficient per unit length is given by the Rayleigh formula

$$\sigma'_\lambda = [32\pi^3(n_\lambda - 1)^2/3N\lambda^4]f \qquad (3.17)$$

where $n_\lambda \simeq 1$ is the refractive index, N is the number of particles per unit volume, and f is the Cabannes correction factor which must be included when the particles are nonspherical [2].

The scattering coefficient for the whole atmosphere may thus be written in the form

$$\sigma_\lambda = [32\pi^3(n_\lambda - 1)^2 H/3N\lambda^4]f \qquad (3.18)$$

TABLE 3.2. Values of scattering coefficient σ_λ' and transmissivity q_λ [3].

Wavelength (μ)	σ_λ' cm^{-1}	q_λ
0.20	954×10^{-8}	0.0005
0.25	338×10^{-8}	0.0669
0.30	152×10^{-8}	0.295
0.35	79×10^{-8}	0.530
0.40	45×10^{-8}	0.696
0.45	28×10^{-8}	0.800
0.50	18×10^{-8}	0.865
0.60	8.6×10^{-8}	0.933
0.70	4.6×10^{-8}	0.964
0.80	2.7×10^{-8}	0.979
0.90	1.7×10^{-8}	0.987
1.00	1.1×10^{-8}	0.991
1.10	7.4×10^{-9}	0.994
1.20	5.3×10^{-9}	0.996
1.50	2.1×10^{-9}	0.998
2.0	6.8×10^{-10}	0.999
4.0	4.2×10^{-10}	1.000

Values of σ'_λ for the atmosphere are given in Table 3.2, which also contains the fraction q_λ of the energy transmitted to the ground through the vertical of a Rayleigh atmosphere. It is clear from Table 3.2 that the scattering coefficient for blue light ($\lambda \sim 0.4\,\mu$) is much larger than for red light ($\lambda \sim 0.7\,\mu$). Indeed, Eq. (3.17) shows that σ'_λ is proportional to λ^{-4}. However, since n_λ is also a function of λ, the combined wavelength dependence is somewhat different from the inverse fourth-power law. Thus, since the sky light consists almost entirely of light scattered from the solar beam, the sky has a blue colour when only air molecules are responsible for scattering. In such a 'pure' atmosphere the sun tends to become redder as the optical air mass increases, while the sky becomes more blue. This effect is due to a greater fraction of the blue component of solar radiation being scattered from the direct solar beam by the greater number of air molecules in the longer path. As blue light is removed from the direct beam, the beam itself appears in consequence to contain a larger proportion of the red component.

In the presence of large particles, however, light of all colours is scattered to a greater extent so that the sky will appear less blue and eventually will become white when a sufficient number of large particles is present, as in the case of a natural cloud of water or ice particles.

It is evident from Table 3.2 that the atmosphere is practically opaque at $\lambda = 0.2\,\mu$, so that even without absorption very little solar energy would reach the ground at this wavelength. The transmissivity increases with λ, but since ozone absorbs very strongly, practically no energy reaches the earth's surface for $\lambda < 0.29\,\mu$. Finally, in a pure Rayleigh atmosphere, almost all the solar energy would reach the ground for $\lambda > 1.0\,\mu$ when the sun is in the zenith. However, scattering by particles and clouds, and absorption (by water vapour and other gases, and particles) often depletes much of the solar energy even at these wavelengths.

The intensity of a direct solar beam transmitted through the atmosphere has been measured for many years by the Smithsonian Institution from mountain peaks. Although the measurements cannot be perfect, approximate agreement with Eq. (3.17) has been achieved. By measuring the scattering coefficient per unit length and the refractive index, it is possible to compute either f or N. It is found that $f = 1.054$ gives the correct result for N [1, 2]. Although some uncertainty may remain about the exact values of the constants, the main features of Rayleigh's law may now be regarded as well established.

Eq. (3.17) gives the scattering coefficient which determines the total amount of energy scattered per unit path length. In the case of primary (as opposed to multiple) scattering, Rayleigh also found that the intensity distribution in the scattered radiation is given by

$$I_\lambda(\theta) \sim I_{0\lambda}(1 + \cos^2\theta)\,(n^2 - 1)^2/\lambda^4 \tag{3.19}$$

where θ is the angle between the direction of incidence and the direction of scattering. Maximum scattering thus occurs in the forward and backward directions. Moreover, there is the very strong dependence on λ (cf. Table 3.2).

Rayleigh also showed that, for a single scatter, the energy scattered at right angles to the solar beam is highly polarized. Fig. 3.3 shows the relative amounts of scattering for green ($\lambda \sim 0.5\,\mu$) and red ($\lambda \sim 0.7\,\mu$) radiation, which is in agreement with Eq. (3.18) when the particles are smaller than λ.

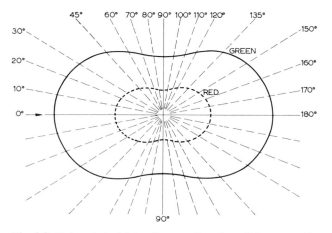

Fig. 3.3. Polar plot of intensity as a function of the scattering angle for small particles ($r \cong 0.025\,\mu$) for green and red light. Wavelengths: green $\cong 0.5\,\mu$, red $\cong 0.7\,\mu$ [4].

The green light is scattered more than the red, but in each case the intensities scattered in the forward and backward directions are equal to twice the intensities at right angles to the incident beam.

In a thick layer of gas the incident radiation may be scattered many times. Some of the energy is scattered in the downward direction giving the blue appearance to the sky, but a considerable proportion is also scattered in the upward direction. Figs. 3.4 and 3.5 show the intensity distribution in upward and downward scattered radiation for different values of the earth's

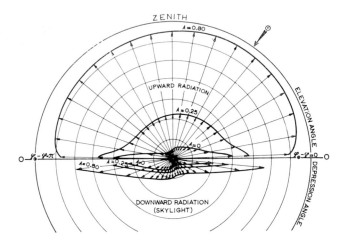

Fig. 3.4. Distribution of the intensity of upward radiation emerging from the top of the atmosphere and of the skylight. The sun's elevation is 53° and its position is indicated by ←○. Moderate optical thickness ($\mu = 0.15$); earth's albedo $A = 0, 0.25, 0.80$ [5].

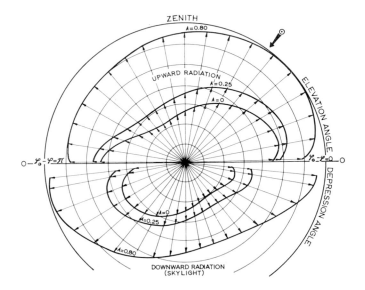

Fig. 3.5. Distribution of the intensity of upward radiation emerging from the top of the atmosphere and of the skylight. The sun's elevation is 53° and its position is indicated by ←○. Large optical thickness ($\mu = 1.0$); earth's albedo $A = 0, 0.25, 0.80$ [5].

albedo* A and two values of the transmissivity, namely, $q_\lambda = 0.86$ and $q_\lambda = 0.37$ ($\mu = 0.15$ and 1.0 respectively, where μ is the optical thickness given by $q_\lambda = \exp(-\mu_\lambda)$).

These figures show the distribution in the plane containing the incident solar beam and the line to the zenith. For a more opaque atmosphere the amount of energy scattered is of course greater, both in the upward and downward directions. Fig. 3.5 also shows that the intensity scattered in the backward direction is somewhat greater than the downward scattered intensity. This is due to multiple scattering. Moreover, these figures show that the scattered intensity, and especially the upward scattered intensity, is a rapidly varying function of the surface albedo A, because the upward intensity includes the energy reflected by the surface. The variation with surface albedo is less rapid for the larger optical thickness, because the intensity reaching the surface in the total direct beam is smaller. Coulson [5] also showed that the intensity of multiply scattered light in the plane of the sun's vertical, for $q_\lambda = 0.37$, may be greater than the intensity of the radiation scattered in primary events, by a factor of three.

The total flux is obtained by integrating over all angles. Fig. 3.6a shows the spectral distribution of the upward scattered flux for several zenith distances. The wavelength corresponding to the maximum flux is seen to increase with the zenith distance. This is also true for the downward scattered radiation as shown in Fig. 3.6b. The entire distribution of the upward scattered energy is indeed similar to the downward scattered energy. It must be recorded, however, that these computations are for a molecular Rayleigh atmosphere and do not take into account effects due to larger particles whose scattering functions are different from those for molecules.

By integrating over all angles it is found that the total energy reflected by a cloudless atmosphere alone is 7% when ozone absorption is taken into account. Measurements of the scattered sky radiation may be compared with computations based on multiple Rayleigh scattering and shown in Fig. 3.6b. For wavelengths greater than about $0.4\,\mu$, where the observed energy is greater than that predicted for a Rayleigh atmosphere, the increase in the downward scattered intensity may be due to large 'Mie particles' (see below). For wavelengths below $0.4\,\mu$ the observed energy is smaller, owing mainly to the absorption by ozone.

---

* The albedo is equal to the ratio of radiation reflected from a surface to that originally falling upon it.

References pp. 107–110

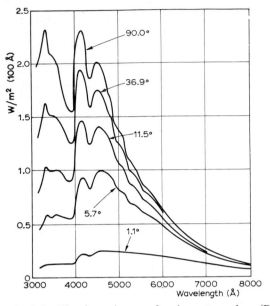

Fig. 3.6a. Flux from the top of a plane atmosphere (Rayleigh scattering only) as a function of wavelength for five values of solar zenith distance $z_\odot$. Zero surface albedo [5].

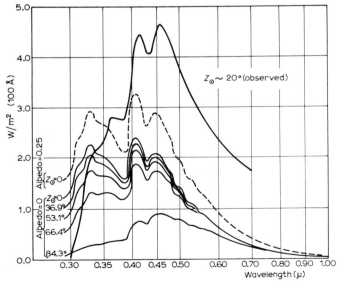

Fig. 3.6b. Flux from the sky alone for Rayleigh scattering. Albedo refers to reflectivity of earth's surface. Curves for various values of zenith distance $z_\odot$ are shown. Curve marked 'observed' is for Cleveland on a clear day in midsummer near noon [7].

When the scattering particles are small in comparison with the wavelength, the Rayleigh scattering theory can be applied. In fact, Rayleigh's theory is applicable when r is roughly speaking $\leqslant 0.1\ \lambda$. When $r \geqslant 25\ \lambda$, geometrical optics may be used to compute the scattering effect. However, in the range $0.1\ \lambda < r < 25\ \lambda$ it is necessary to use the more complicated theory developed by Mie [1].

The scattering by spherical particles is due to the interaction between light waves which have been reflected, refracted and diffracted by the particles. The diffraction effects are only slight when the particles are very large. In general, the total scattering cross-section s is given by

$$s = \pi r^2\, K(y, n) \tag{3.20}$$

where K is a dimensionless function by which the geometrical cross-section must be multiplied in order to obtain the total scattering cross-section. The magnitude of $K(y, n)$ depends on the parameter $y = 2\pi r/\lambda$ and on the refractive index n. Fig. 3.7 shows typical curves of K for three values of n. These curves are based on approximate calculations carried out by Penndorf [6].

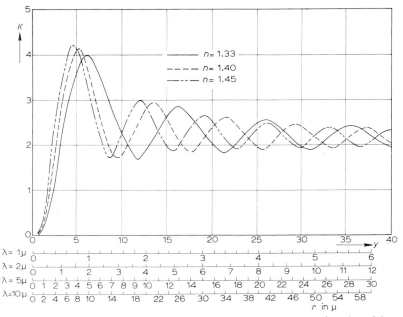

Fig. 3.7. Smoothed values of the Mie scattering cross-section K as a function of the parameter y, for three values of the refractive index n. Dependence of K on radius for fixed values of λ are also shown by scales below the figure [6].

Penndorf has computed approximations to the Mie cross-section to an estimated error of less than 3%. Detailed calculations have shown that these curves exhibit considerable fine structure (Fig. 3.8). The maxima of K in Fig. 3.7 occur at progressively smaller values of y as the refractive index increases. Hence for a given λ, a given maximum occurs at progressively smaller particle sizes as the magnitude of the refractive n increases. Similarly, for a given r, a given maximum of K occurs at progressively larger values of λ as the refractive index increases.

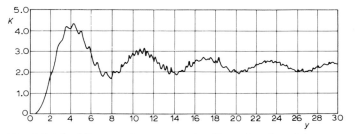

Fig. 3.8. The Mie scattering cross-section K as a function of the parameter y for $n = 1.50$ [6].

Fig. 3.7 also shows that the scattering cross-section may exceed the geometric cross-section by a factor of 4 when the wavelength approaches the particle radius ($n \sim 1.33$, i.e. water). It is clear from Fig. 3.7 that K is an oscillatory function with a decreasing amplitude approaching the value of 2 when y is large. This indicates that the scattering cross-section approaches twice the geometrical cross-section for large spherical particles. For other values of n the curves are similar in form.

Whereas Rayleigh scattering is symmetrical with respect to the direction of incidence, the angular scattering by Mie particles depends on the parameter y, and the amount of light scattered in the forward is much larger than in the backward direction. This is illustrated by the angular distribution shown in Fig. 3.9.

For very large drops, Wiener's calculations have been used. Wiener's distribution is rather similar to that discussed below in connection with Fig. 3.10.

In a mixture of particles exposed to sunlight the magnitude of y will vary throughout the spectrum. Moreover, y will vary with particle size at given λ. Computations must, therefore, involve integrals over both λ and r.

Fraser has collected some experimental results and compared them

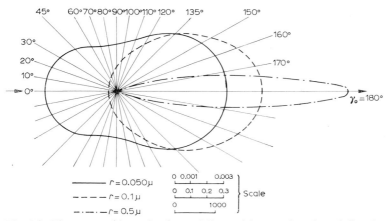

Fig. 3.9. The scattered intensity due to Mie particles as a function of direction for green light [1].

with his calculations for a mixture of particles at a single wavelength. These results are shown in Fig. 3.10 which is a compilation of intensity measurements made with searchlight beams by Reeger and Siedentopf (RS) and by Siedentopf (S).

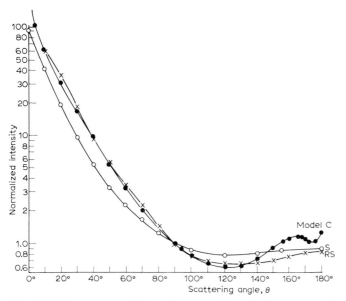

Fig. 3.10. Mean measured intensity on continental sites compared with computed intensity for $\lambda = 0.625 \mu$ [8].

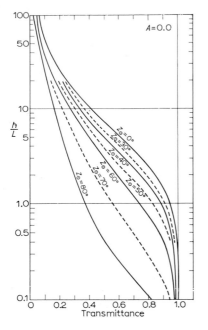

Fig. 3.11. Transmittance through overcast cloud, *i.e.* fraction of direct beam incident on horizontal cloud top which is transmitted to ground. The ordinates represent the cloud's optical thickness. The sun's zenith distance $z_\odot$ is as indicated on the curves; surface albedo $A = 0$ [7].

The curve labelled 'model C' is based on the Mie theory [8], using an aerosol which is typical of Continental Europe. Both measurements and computations clearly show that the forward intensity is larger than the intensity at about 120°, by a factor greater than 100. The intensity in the backward direction (180°) is still quite small but is somewhat larger than the minimum near 120°. In model C there were about 1500 particles per cubic centimetre with radii in the range 0.03–0.1 μ and about 750 particles per cubic centimetre with radii in the range 0.1–20 μ.

The problem of multiple scattering in dense aerosols consisting of large particles has not been considered as extensively as in the case of Rayleigh scattering. Typical transmittance curves for relatively large water particles and zero surface albedo are shown in Fig. 3.11 in which h is the vertical thickness of the cloud, and the mean free path L is given by

$$L = (\sum_i N_i \pi r_i^2)^{-1} = \tfrac{4}{3}(r_e/w) \qquad (3.21)$$

where N_i is the number of particles of radius r per unit volume, r_e is the effective scattering radius and w the liquid water contained in a unit volume. The quantity h/L is therefore analogous to an effective optical thickness of the cloud. In Fig. 3.11 it is assumed that no absorption occurs in the cloud.

It is evident from Fig. 3.11 that for a given zenith distance $z_{\odot}$, the transmittance decreases and therefore the albedo increases with h/L. For thick clouds the albedo approaches unity for all values of $z_{\odot}$ (the cloud albedo is the fraction of incident light reflected by the cloud top into space). However, for h/L of the order of unity (moderately thin clouds), the albedo is a rapidly varying function of $z_{\odot}$ and can approximately be represented by $\sec z_{\odot}$.

The transmitted energy and the energy reflected back into space depends on the surface albedo. The dependence of the transmitted energy on the surface albedo is shown in Fig. 3.12 which gives the fraction of the transmitted energy T as a function of h/L for $z_{\odot}=0$. The dependence of T on the surface albedo (reflectance) is also shown.

The luminance of the sky which is due to scattering depends on the position of the sun and the coordinates of the particular point on the celestial sphere. Dorno [10] has carried out extensive measurements and mappings of

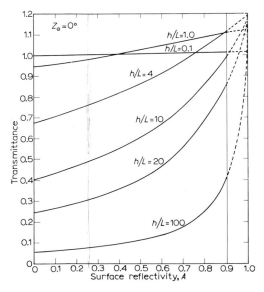

Fig. 3.12. Transmittance through clouds as a function of surface albedo A, for several values of h/L; zero zenith distance [9].

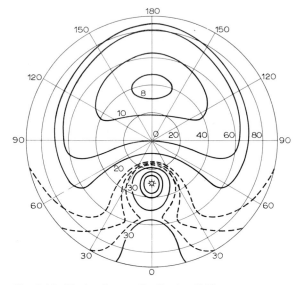

Fig. 3.13. Sky luminance distribution [10].

TABLE 3.3. Sky luminance (in relative units) for Davos [10].

Sun	Point on sky	Azimuth of point observed		
		0	90	180
0	20	10.68	3.04	3.98
	30	5.86	2.46	2.98
	45	2.55	1.56	4.71
	60	1.51	1.31	1.26
20	20	—	2.58	3.05
	30	6.16	1.71	2.03
	45	3.35	1.34	1.37
	60	1.87	1.20	1.01
40	20	6.30	2.03	1.82
	30	5.60	1.67	1.34
	45	—	1.15	0.92
	60	2.35	1.06	0.73
60	20	2.10	1.22	0.72
	30	2.10	0.88	0.62
	45	2.35	0.80	0.50
	60	—	0.81	0.52

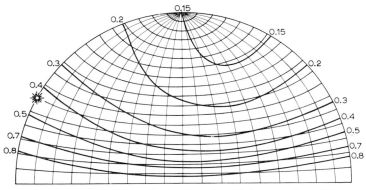

Fig. 3.14. Sky luminance distribution for a Rayleigh atmosphere [13].

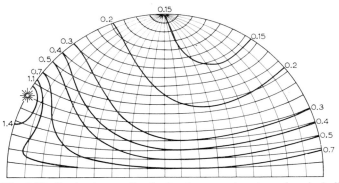

Fig. 3.15. Sky luminance distribution for a regular atmosphere including multiple scattering [13].

the sky luminance at Davos (Switzerland). Some typical results obtained by him are illustrated in Fig. 3.13 and Table 3.3. The luminance can be expressed either in absolute units (stilbs) or in relative units with the luminance at the zenith taken as unity (as in the data of Table 3.3). Other valuable information on illumination is given by Veynberg [11].

Fig. 3.14 shows the results of luminance calculations carried out by Pyaskovskaya-Fesenkova for a Rayleigh atmosphere with multiple scattering and Fig. 3.15 for a regular atmosphere with multiple scattering. The results are expressed in stilbs with $\gamma_{\odot} = 60°$ and $q = 0.861$ for both figures.

The luminance distribution over the sky, which may be obtained experimentally or by calculation, is of particular importance in architectural design [12].

The spectral composition of the radiation will be discussed in Chapter 5.

4. ABSORPTION IN THE ULTRAVIOLET

Whereas atmospheric scattering is a continuous function of the wavelength, atmospheric absorption is in general selective. The main absorbing components of the atmosphere are O_2, O_3, H_2O, CO_2, N_2, O, N, NO, N_2O, CO, CH_4 and their isotopic modifications, although the contributions of the latter are small.

Spectra due to electronic transitions of molecular and atomic oxygen and nitrogen, and of ozone, lie chiefly in the ultraviolet region, while those due to the vibration and rotation of polyatomic molecules such as H_2O, CO_2 and O_3 lie in the infrared. There is very little absorption in the visible region.

As the absorption coefficients associated with electronic transitions are generally very large, much of the solar ultraviolet radiation is absorbed in the upper layers of the atmosphere. Some of the oxygen and nitrogen molecules are dissociated into atomic oxygen and nitrogen owing to the absorption of solar radiation, while other molecules become ionized. Dissociated atomic oxygen and nitrogen are also able to absorb solar radiation of still shorter wavelength and some of these atoms become ionized as a result. The ionized layers in the upper atmosphere are formed mainly as a consequence of these processes. Some of the absorbed energy is re-emitted by the excited atoms and molecules as they relax to their ground states, while the remainder is used to heat the particular layer.

Ozone, which exists in the upper part of the stratosphere, is also formed as a result of the photochemical action of solar radiation.

Owing to the very strong absorption by O_2, N_2, O, N and O_3 in the spectral region up to about 3000 Å, the solar radiation in this region does not reach the earth's surface.

In the visible region there is some absorption due to the weak Chappuis bands of ozone and to the red bands of molecular oxygen which occur at about 0.69 and 0.76 μ.

In the infrared, absorption by water vapour occurs at about 0.7, 0.8, 0.9, 1.1, 1.4, 1.9, 2.7, 3.2 and 6.3 μ and by carbon dioxide at about 1.6, 2.0, 2.7 and 4.3 μ. These bands play a part in the absorption in the lower atmosphere, below 50 km, where water vapour and carbon dioxide are largely concentrated. No photochemical action is associated with absorption in this region and the absorbed energy is used entirely to heat the lower atmosphere.

Beyond the 6.3 μ bands of H_2O there is an almost transparent region, the so-called atmospheric window, which lies between 8 and $12\,\mu$. In the range between $13\,\mu$ and wavelengths of the order of millimetres, solar radiation is almost completely absorbed by the strong rotational bands of H_2O and the $15\,\mu\,CO_2$ bands.

In the millimetre and centimetre ranges the absorption is caused by several rotational lines of H_2O and also by several O_2 lines.

Finally, it should be noted that the earth's surface is generally a good absorber, so that considerable amounts of solar radiation are absorbed by it in the visible and in the near infrared.

It will perhaps be useful at this stage to introduce the definition of the absorption coefficient again. For a beam of monochromatic radiation of frequency v and intensity I_v falling perpendicularly on the surface of an absorbing layer of thickness dx, the attenuation of the incident radiation by the layer is described by

$$-dI_v = I_v s_v n\,dx$$

where n is the number of absorbing particles per unit volume and s_v is the absorption cross-section, which is often used in the ultraviolet region. Moreover, we can write

$$n\,dx = n_0\,dx_0$$

where n_0 is the Loschmidt number (2.687×10^{19} cm^{-3}) and dx_0 is the thickness at normal temperature and pressure. The quantity $\alpha'_v = s_v n_0$ is usually referred to as the absorption coefficient.

A further quantity which is often used is the so-called mass absorption coefficient which is equal to $s_v n/\varrho$.

The ultraviolet spectrum of molecular oxygen begins at about 2600 Å and continues down to shorter wavelengths. The bands lying between 2600 and 2000 Å, frequently referred to as the Herzberg bands, are very weak and of little importance in the absorption of solar radiation, owing to the presence of the much stronger ozone bands in this spectral region. The Herzberg bands are however considered to be of importance in the formation of ozone. Adjacent to the Herzberg bands are the very strong Schumann–Runge bands which begin at 2000 Å and continue down to about 1250 Å.

Several bands exist between 1250 and 1000 Å. Most of them are members of the Rydberg series converging to the first ionization potential at 12.08 eV (1026.5 Å), although the observed bands have not yet been definitely identified. The region below 1000 Å is occupied by the very strong

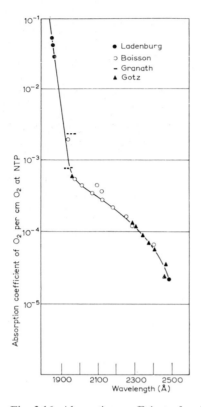

Fig. 3.16. Absorption coefficient of molecular oxygen in the Herzberg bands [14].

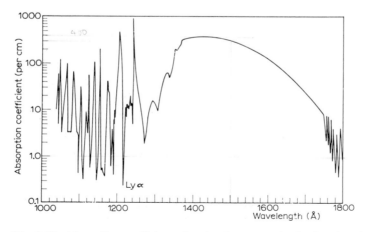

Fig. 3.17. Absorption coefficient of molecular oxygen in the far ultraviolet [15].

O_2 bands which are referred to as the Hopfield bands. The latter consist of three Rydberg series converging to the second, third and fourth ionization potentials at 16.1, 16.9 and 18.2 eV (770, 730 and 680 Å respectively).

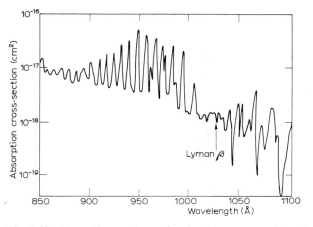

Fig. 3.18. Absorption continua of molecular oxygen and the Hopfield bands in the region 850–1100 Å [16].

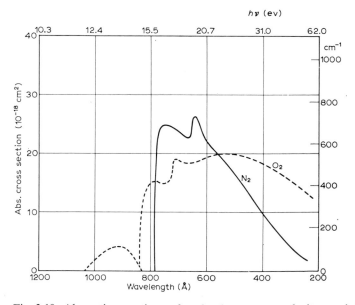

Fig. 3.19. Absorption continua of molecular oxygen and nitrogen below 1000 Å [17].

The absorption coefficients of O_2 in the ultraviolet have been measured by many workers. Those compiled by Craig [14] for the region containing the Herzberg bands are shown in Fig. 3.16, while those measured by Watanabe et al. [15] in the region 1800–1000 Å are shown in Fig. 3.17. Measurements by Watanabe and Marmo [16] for the 1100–850 Å region are given in Fig. 3.18, and those reported by Lee [17] for the region below 1000 Å are shown in Fig. 3.19.

As can be seen from Fig. 3.17 the Lyman α-line, which is very strong in the solar spectrum, lies in one of the 'windows' of the O_2 absorption spectrum.

The absorption spectrum of molecular nitrogen begins at 1450 Å and the bands occupying the region 1450–1000 Å are called the Lyman–Birge–Hopfield bands. They consist of narrow and sharp lines between 1450 and 1130.4 Å and possibly include the band at 1114.2 Å. Although the absorption spectrum of N_2 in the region between 1000 and 800 Å is very complex, most of the prominent bands in this region have been identified as members of Rydberg series converging to the first ionization potential at 15.58 eV. In the region below 800 Å the spectrum is in general made up of an ionization continuum with members of Rydberg series superimposed on it down to 650 Å.

The absorption cross-sections corresponding to the Lyman–Birge–Hopfield bands are very small. According to Watanabe [18] they are 7×10^{-21} cm^2 at the peaks of the bands and 3×10^{-22} cm^2 between the L–B–H bands. Ditchburn et al. [19] have reported an even lower value of 6×10^{-23} cm^2 for the minimum cross-section between the L–B–H bands.

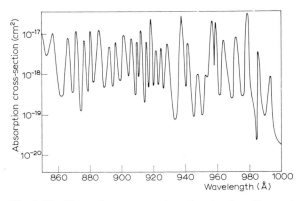

Fig. 3.20. Absorption cross-section of molecular nitrogen in the region 850–1000 Å [16].

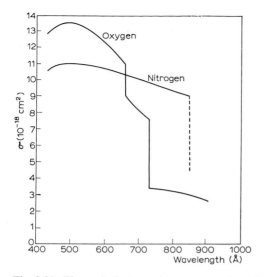

Fig. 3.21. Theoretical absorption cross-section of atomic oxygen and nitrogen in the extreme ultraviolet [20].

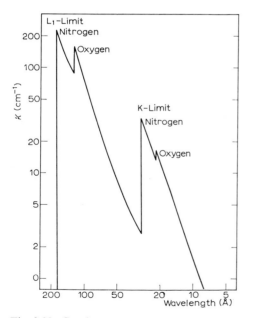

Fig. 3.22. Continuous X-ray absorption of atomic nitrogen and oxygen [21].

The absorption cross-sections in the region 1000–850 Å which have been measured by Watanabe and Marmo [16] are shown in Fig. 3.20, and those in the region below 800 Å (Lee) are shown in Fig. 3.19 [17].

The presence of the strong Schumann–Runge continuum in the ultra-violet spectrum of O_2 (Fig. 3.17) proves that a portion of molecular oxygen in the upper atmosphere is dissociated into atomic oxygen by solar radiation. Photo-dissociation of nitrogen molecules is also expected in the upper atmosphere. It follows that the absorption cross-sections of O and N are also of importance, although so far they have only been evaluated from quantum mechanical calculations. Fig. 3.21 shows the results obtained in this way by Bates and Seaton [20].

At wavelengths below about 200 Å, as one approaches the X-ray region, the absorption cross-sections of O and N have been estimated by Nicolet [21] and are shown in Fig. 3.22.

Ozone exhibits weak absorption bands in the near infrared and in the visible (11,800–4400 Å) regions, known as Chappuis bands. The ozone bands between 3600 and 3000 Å are called Huggins bands. The strongest bands of ozone, which lie directly below the Huggins bands and extend from 3000 to 2000 Å, are called Hartley bands. Fig. 3.23 shows the absorption coefficients of ozone. In Figures 3.23b and c the dots represent the values obtained by Vigroux, while the solid curves represent the values by Inn and Tanaka. The values obtained by Vigroux are the ones used most frequently by ozone researchers.

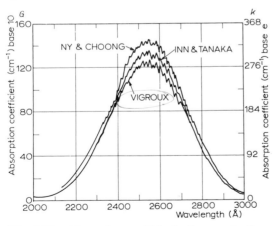

Fig. 3.23a. Absorption coefficient of ozone in the region 2000–3000 Å (*Handbook of Geophysics*, Revised edition, U.S. Air Force, Macmillan, New York, 1960; cf. [22–24]).

In the region below 2000 Å the absorption by ozone was formerly believed to be very weak. Recently, however, Tanaka *et al.* [25] have succeeded in measuring the absorption coefficients of ozone in the region between 2200 and 1050 Å; these results are shown in Fig. 3.24.

The data on the absorption cross-sections of atmospheric gases and the composition and physical nature of the upper atmosphere have been used by a number of workers to calculate the penetration of solar radiation into

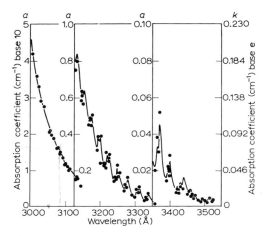

Fig. 3.23b. Absorption coefficient of ozone in the region 3000–3500 Å. Dots: Vigroux [23]; solid curves: Inn and Tanaka [24].

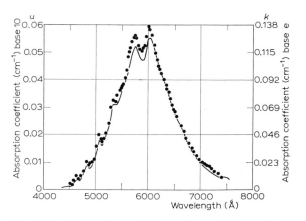

Fig. 3.23c. Absorption coefficient of ozone in the region 4000–8000 Å. Dots: Vigroux [23]; solid curves: Inn and Tanaka [24].

References pp. 107–110

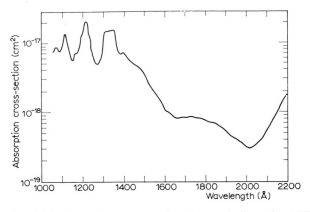

Fig. 3.24. Absorption cross-section of ozone in the region 1050–2200 Å [25].

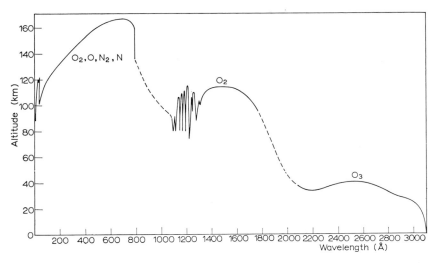

Fig. 3.25. Penetration of solar radiation into the atmosphere. The curve indicates the level at which the intensity is reduced to e^{-1} [26].

the atmosphere. The estimates obtained by Friedman [26] are shown in Fig. 3.25 which gives the altitude at which the intensity of solar radiation at vertical incidence is reduced by a factor of e. This figure will serve to indicate the general nature of the variation of the atmospheric transparency with wavelength. In the region 3000–2000 Å absorption is due to ozone, between 2000 and 850 Å it is due to molecular oxygen, while below 850 Å the components chiefly responsible for absorption are O_2, O, N_2 and N.

5. FORMATION OF IONOSPHERIC AND OZONE LAYERS

The absorption of solar ultraviolet radiation is closely associated with photochemical reactions which lead to the formation of ionized layers. The formation theory was first proposed by Chapman [27] and was later developed by many other workers. No conclusive theory exists however at the present time, and in order to verify or improve the theoretical data more precise experimental results are necessary on the distribution of the atmospheric constituents, the intensity of solar ultraviolet radiation and on the absorption cross-sections of atmospheric gases. A brief outline of the theory of formation of ionized layers is given below.

The ionization of air begins to increase at about 60 km above sea level and the first layer, which occupies the region between 60 and 90 km, is called the D-layer. According to Nicolet [28], the normal D-layer is produced as a result of the reaction $NO + h\nu \rightarrow NO^+ + e$, where the incident photon energy is 9.25–12 eV. The necessary radiation is ascribed mainly to the solar Lyman α-line. It is clear from the above discussion and from Fig. 3.25, that this radiation (1215.7 Å) is capable of penetrating to about 70 km, *i.e.* into the D-layer. The enhanced ionization in the D-layer during radio fadeout has been ascribed to an increase in the intensity of the Lyman α-line and solar X rays below 10 Å. According to Friedman [29] the latter appears to be the primary cause.

Two theories have been proposed as to the origin of the E-layer which lies above the D-layer at an altitude of about 100 km.

One of them was proposed by Wulf and Deming [30] and was later developed by Nicolet [28], who suggested that the strong Hopfield bands in the region 1027–800 Å might be pre-ionized and that the E-layer may be produced by the two-step reaction $O_2 + h\nu$ (Hopfield bands) $\rightarrow O_2^* \rightarrow O_2^+ + e$ where the asterisk indicates an excited state.

The other theory is that originally proposed by Hulbert [31] and Vegard [32] and later improved by Hoyle and Bates [33]. According to this interpretation, X rays of about 40 Å from the solar corona should produce maximum ionization in the E-layer. Strong evidence in favour of the X-ray theory has recently been obtained from rocket experiments. According to the rocket measurements of Byram *et al.* [34], the total solar flux in the X-ray region is 0.1 erg $\cdot$ cm^{-2} $\cdot$ sec^{-1} which is sufficient to account for the entire ionization observed in the E-layer.

The mechanism responsible for the formation of the F-layer is extremely complicated and no satisfactory theory is available at present.

There is no essential difference between the formation of the ozone layer and the ionized layers: they are both due to solar photochemical reactions. The work in this field is again due to Chapman [35]. The reaction involved in the formation of the ozone layer is $O_2 + hv \rightarrow O + O$. Both the intense Schumann–Runge continuum and the weak Herzberg continuum participate in this reaction. However, the photo-dissociation due to the Schumann–Runge continuum is confined to high altitudes because of the large cross-section. In contrast, the photo-dissociation associated with the Herzberg continuum is significant throughout the atmosphere.

The oxygen atoms produced in the above reaction may be removed by any of the following reactions: $O + O + M \rightarrow O_2 + M$, $O + O \rightarrow O_2 + hv$, $O + O_2 + M \rightarrow O_3 + M$. The symbol M in these formulae represents a third atom or molecule whose presence is necessary in order to conserve energy and momentum. The ozone produced in the last of these three reactions may disappear as a result of $O_3 + O \rightarrow 2\,O_2$ or $O_3 + hv \rightarrow O_2 + O$. Many workers assume equilibrium conditions and use the above reactions to calculate the distribution of ozone. The reaction $O + O \rightarrow O_2 + hv$ used to be regarded as ineffective until Bates and Nicolet [36] first took it into consideration and Johnson *et al.* [37] emphasized its importance in the region above 60 km, where three-body recombination processes begin to decrease.

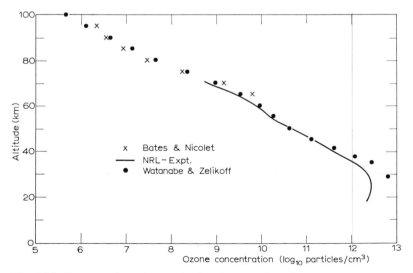

Fig. 3.26. Concentration of ozone in decadic logarithms (particles/cm³) as a function of altitude [36–38].

Fig. 3.26 shows the results of calculations by Bates and Nicolet [36] and the more recent results of Watanabe and Zelikoff [38] together with the rocket results of Johnson *et al.* [37]. It will be observed that there is good agreement between experiment and theory.

The vertical distribution of ozone and the total amount of ozone in a vertical column have been determined by many workers. Comparisons of these observations with the vertical distribution predicted by the photo-chemical equilibrium theory has shown that the latter holds only above 35 km. Below 35 km the equilibrium theory fails to explain the observed ozone profiles. More ozone is predicted by the theory at the level of the ozone layer maximum which occurs at about 25–30 km than is indicated by the observations. On the other hand the amount of ozone predicted by the theory for the lower altitudes is found to lie below experimental results.

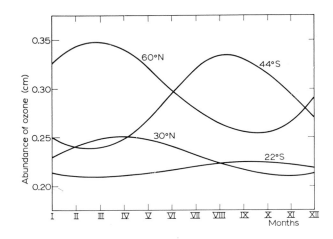

Fig. 3.27. Seasonal and latitudinal variation of the ozone content in a vertical air column [39, 40].

Seasonal and latitudinal variations in the total amount of ozone in the vertical air column have also been extensively studied. The nature of the variation was reviewed by Bates [39] and is shown in Fig. 3.27 which is based on data reported by Dobson [40]. It is interesting to note that the maximum amount of ozone occurs in the spring and the minimum in the autumn and that both the absolute value and the fractional change are less at the low latitudes. The amount of ozone is also found to depend on atmospheric patterns. For example, it was shown by Dobson *et al.* [41] and Tønsberg and

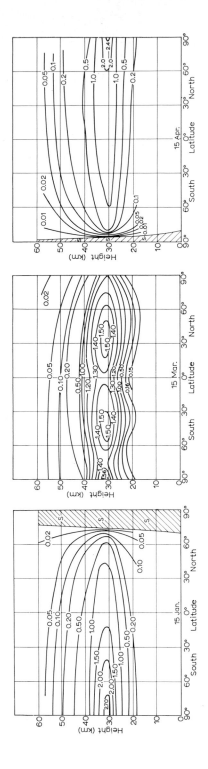
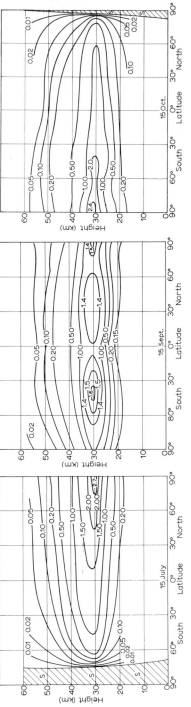

Fig. 3.28. Absorption isopleths of solar radiation by ozone for six different dates [44].

Olsen [42] that in certain cyclones a rapid rise in the amount of ozone follows the passage of a cold front.

The detailed observational results have not yet been fully explained. As noted above, the equilibrium theory has failed to explain even the vertical distribution. Wulf and Deming [43] have also examined the rate at which the equilibrium state is approached and found that whereas the equilibrium state is reached rapidly at altitudes above 35 km, longer periods are required in the lower layers. In the lower layers the global circulation and mixing are important in the explanation of the ozone distribution.

Pressman [44] has calculated the seasonal and latitudinal variations as well as the vertical distribution of the absorption of solar radiation by ozone. Fig. 3.28 shows some of his results on ozone solar absorption; S denotes the shadow zone where no absorption takes place.

In January a single absorption maximum of 2.7 cal $\cdot$ km^{-1} $\cdot$ cm^{-2} $\cdot$ day^{-1} occurs over the south pole and gradually decreases to zero at the shadow zone at about 67°N. A similar pattern continues throughout February. In March two maxima of 1.5 cal $\cdot$ km^{-1} $\cdot$ cm^{-2} $\cdot$ day^{-1} appear at roughly 40°N and 40°S while a third maximum of equal magnitude is found over the south pole. This change in absorption is due to the fact that the sun has moved forward close to, but as yet not right up to, the equator. At the south pole the length-of-day factor creates a maximum, whereas this factor together with the latitudinal distribution of ozone are responsible for the other two maxima. In April only one maximum appears, this time at the north pole and a similar pattern continues until September when it resembles the March pattern except that the third maximum occurs over the north pole. From October until December a single maximum appears again over the south pole.

6. ABSORPTION IN THE INFRARED

Absorption bands in the infrared region are chiefly due to vibrational and rotational transitions of polyatomic molecules, except for the oxygen bands observed in the solar spectrum; for example, the A and B bands in the red region of the solar spectrum. The A band is particularly well known because it led to the discovery of the isotopes ^{18}O and ^{17}O. Kieke and Babcock [45] have found an extremely weak A′ band in the solar spectrum near the intense A band at 7596 Å which was identified as an ^{18}O^{16}O band by

Giauque and Johnston [46]. Shortly after the discovery of the ^{18}O isotope Babcock discovered another band (A'') which was identified as an $^{17}O^{16}O$ band. Other O_2 bands exist in the near infrared, the strongest occurring at 12,680 Å.

Molecular bands between 0.3 and 24 μ which take part in the absorption of solar radiation were compiled by Goldberg [47]. Detailed solar spectra, which indicate spectral lines of solar and telluric origin, have been given by Babcock and Moore [48] (6600–13,495 Å), Mohler *et al.* [49] (0.8–2.5 μ), Shaw *et al.* [50] (3.0–5.2 μ), Shaw *et al.* [51] (7–13 μ) and Migeotte *et al.* [52] (2.8–23.7 μ).

The solar spectrum recorded with a low resolution spectrometer is useful in obtaining a general view of solar infrared absorption. An example of such a spectrum is shown in Fig. 3.29 in which the shaded areas represent absorption, and the difference between the solar intensity curves outside the atmosphere and at sea level is due to scattering. Fig. 3.30 which was given by Howard *et al.* [53] shows regions of strong absorption by molecules present in the earth's atmosphere together with a typical near-infrared solar spectrum which was obtained by superimposing the contributions due to the seven constituents. These figures show clearly that the principal regions of absorption which affect the energy reaching the surface are due to H_2O, CO_2 and O_3.

The estimation of the transmittance in the infrared bands is a complicated problem because of the following facts. A band consists of many lines of different intensity, and lines belonging to different vibrational states are frequently superimposed within the same spectral region. Moreover, it is known that each absorption line has a finite width due to natural damping, Doppler effect and a collision effect. Below 50 km the width is chiefly due to the collision effect [54] and the line shape is approximately given by the Lorentz formula

$$\alpha'_\nu = \frac{S}{\pi} \frac{\gamma}{(\nu - \nu_0)^2 + \gamma^2} \tag{3.22}$$

where α'_ν is the absorption coefficient at frequency ν, ν_0 is the frequency at the centre of the line, S is the line intensity given by

$$S = \int_{-\infty}^{+\infty} \alpha'_\nu d\nu \tag{3.23}$$

and γ is the line half-width at half-height. It was shown by Ladenburg and

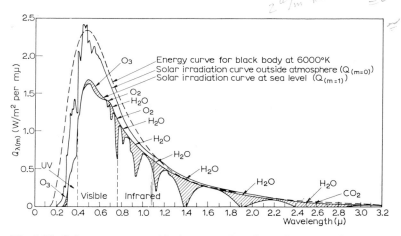

Fig. 3.29. Solar spectrum outside the atmosphere ($m = 0$) compared with that of a black body at 6000°K and at sea level ($m = 1$) (*Handbook of Geophysics*, Revised edition, U.S. Air Force, Macmillan, New York, 1960).

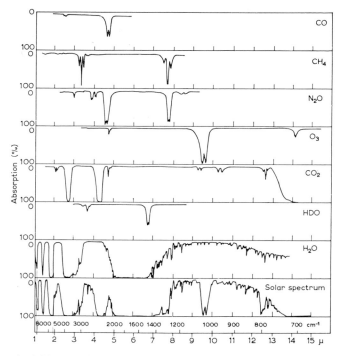

Fig. 3.30. Near-infrared solar spectrum compared with the absorption spectra of CO, CH_4, N_2O, O_3, CO_2, HDO and H_2O [53].

References pp. 107–110

Reiche [55] that the mean transmittance of the line does not follow the Beer–Bouguer–Lambert law. The problem is further complicated due to the fact that the line intensity is a function of the temperature and the line half-width is a function of temperature and pressure.

In consideration of these factors sophisticated methods of calculation of the band transmittance were developed by Elsasser [56], Goody [57] and other workers in this field [58–60]. These methods have been applied successfully to the radiative transfer problem in the far infrared beyond the 6.3 μ H_2O band. However, the H_2O and CO_2 band intensities and line half-widths in the near infrared are still not known with sufficient accuracy to apply these methods. Therefore the estimation of the absorption of solar radiation has been made based on experimental observations.

Stull et $al.$ [61] evaluated the band and line intensities of the main H_2O and CO_2 bands in the near- and far-infrared regions and calculated the transmittance.

Fractional absorption in the main near-infrared bands was determined by Fowle [60] and Howard et $al.$ [53]. Burch et $al.$ [62] measured the absorption of the CO_2 bands and confirmed the results of Howard et $al.$

Although most calculations of the absorption of near-infrared solar radiation have been based on Fowle's absorption curve, we shall present here the more recent and reliable measurements of Howard et $al.$ These laboratory measurements were carried out with a low resolution spectrometer in order to obtain reliable data on the total absorption in each band. Howard et $al.$ used N_2 as the foreign gas to alter the pressure in the cell.

According to Benedict [63] the broadening of the lines due to H_2O— H_2O collisions is greater than that due to N_2—H_2O collisions, by a factor of 6.3.

In the case of CO_2 the broadening due to CO_2—CO_2 collisions is, according to Kostkowski [64], larger than that due to N_2—CO_2 collisions by only 30%. It will be reasonable to take these facts into consideration in presenting the absorption data of Howard et $al.$

Figs. 3.31 a–l show the absorption curves compiled by Yamamoto, based largely on the data of Howard et $al.$, corresponding to the near-infrared bands of H_2O and CO_2 which are responsible for the absorption of solar infrared radiation. For H_2O bands $(p+5.3p_w)/760$ is taken as the pressure parameter where p is the total pressure and p_w the partial pressure of H_2O (both in mm Hg). In the case of CO_2 bands the quantity $p/760$ is taken as the pressure parameter and the 30% self-broadening effect is

(Continued on p. 88)

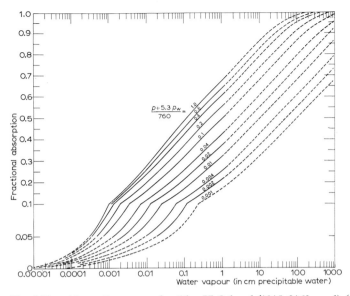

Fig. 3.31a. Absorption curves for 6.3 μ H_2O band (1015–2160 cm^{-1}) (G. Yamamoto; cf. [53]).

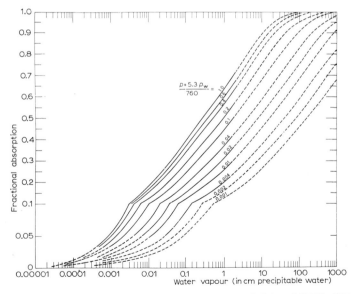

Fig. 3.31b. Absorption curves for 2.7 μ + 3.2 μ H_2O band (2800–4400 cm^{-1}) (G. Yamamoto; cf. [53]).

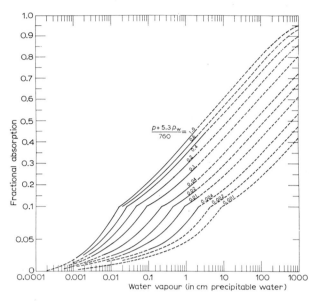

Fig. 3.31c. Absorption curves for 1.87 μ H_2O band (4800–6200 cm^{-1}) (G. Yamamoto; cf. [53]).

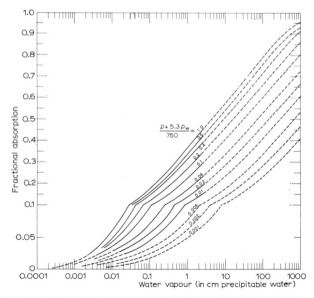

Fig. 3.31d. Absorption curves for 1.38 μ H_2O band (6200–8200 cm^{-1}) (G. Yamamoto; cf. [53]).

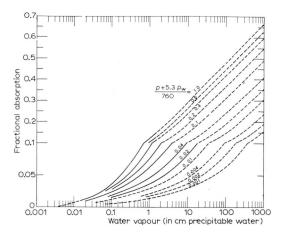

Fig. 3.31e. Absorption curves for 1.1 μ H_2O band (8200–9700 cm^{-1}) (G. Yamamoto; cf. [53]).

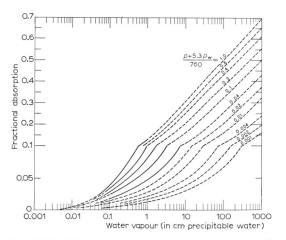

Fig. 3.31f. Absorption curves for 0.94 μ H_2O band (9700–11500 cm^{-1}) (G. Yamamoto; cf. [53]).

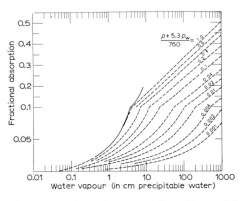

Fig. 3.31g. Absorption curves for 0.81 μ H_2O band (11905–12658 cm^{-1}) (G. Yamamoto; cf. [60]).

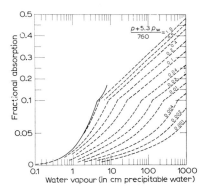

Fig. 3.31h. Absorption curves for 0.72 μ H_2O band (14286–13514 cm^{-1}) (G. Yamamoto; cf. [60]).

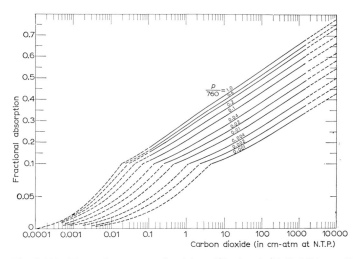

Fig. 3.31i. Absorption curves for 4.3 μ CO_2 band (2160–2500 cm^{-1}) (G. Yamamoto; cf. [53]).

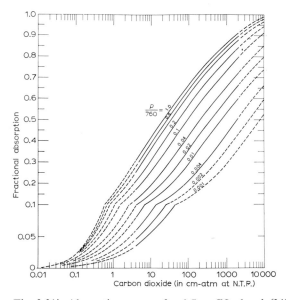

Fig. 3.31j. Absorption curves for 2.7 μ CO_2 band (3480–3800 cm^{-1}) (G. Yamamoto; cf. [53]).

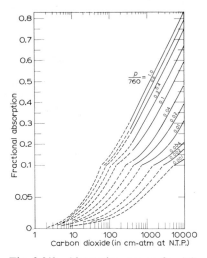

Fig. 3.31k. Absorption curves for 2.0 μ CO_2 band (4750–5050 cm^{-1}) (G. Yamamoto; cf. [53]).

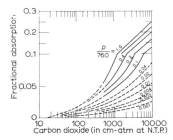

Fig. 3.31l. Absorption curves for 1.6 μ CO_2 band (6000–6550 cm^{-1}) (G. Yamamoto; cf. [53]).

neglected. In the figures the solid lines correspond to the actually observed values and the broken lines represent extrapolations. Since Howard et al. did not measure the absorption in the 0.72 and 0.8 μ bands of H_2O, the curves shown in Figs. 3.31 g and h were compiled on the basis of Fowle's data [60] which are shown by the solid lines.

In order to calculate the absorption of solar radiation with the aid of the laboratory absorption data shown in Figs. 3.31 it is necessary to use a pressure parameter which takes into account the vertical distribution of the absorbing gas in the atmosphere. For instance, for water vapour this parameter is given by

$$P_e = (760 \ u)^{-1} \int_0^\mu (p + 5.3 \ p_w) \mathrm{d}u \tag{3.24}$$

where u is the optical thickness of water vapour measured from the top of the atmosphere downward. The energy absorbed by the i-th band during the day-time from the top of the atmosphere down to a given level $\bar{A}_i$ is given by

$$\bar{A}_i = 2 \int_0^{h_0} I_{0i} A_i(u \ \sec \zeta \ P_e) \cos \zeta \, \mathrm{d}h \tag{3.25}$$

where A_i is the instantaneous value of the absorption of the i-th band, I_{0i} the solar energy outside the atmosphere in the i-th band, ζ the solar zenith angle, h the hour angle, and h_0 the hour angle at sunrise or sunset (these may be assumed to be equal).

Calculated absorptivities are shown in Fig. 3.32 (Yamamoto [65]). The absorptivity is defined as the ratio of the energy absorbed by the vertical air column at normal incidence to the solar constant. Similar calculations have been reported by many workers and most of them are based on Fowle's data (Kimball [66]; Mügge and Möller [67]; Tanck [68]; Yamamoto and Onishi [69]; Houghton [70]; McDonald [71]) although the results were not always in agreement.

The recent determination by McDonald seems to be satisfactory. McDonald's results are indicated by the crosses in Fig. 3.32. In this figure curve 1 shows the absorptivity corresponding to the 0.72, 0.81, 0.94, 1.10, 1.38 and 1.87 μ bands. As can be seen, McDonald's values for the absorptivity are in fair agreement with curve 1. This indicates not only the soundness of Fowle's data which were obtained about 50 years ago but also the soundness of McDonald's interpretation of these data and the results of Howard et al. Curve 2 in Fig. 3.32 shows the absorptivity due to all the H_2O bands from 0.72 to 6.3 μ and this is greater than McDonald's absorptivity owing to the strong absorption at 2.7, 3.2 and 6.3 μ. Finally, the absorptivity due to the sum of the H_2O and CO_2 bands is shown by curve 3, and that due to the sum of the H_2O, CO_2 and O_2 bands by curve 4. The absorption due to O_2 bands was estimated from the data of Fowle [60], Langley and Abbot [72], and Gates [73].

The vertical distribution of heating due to absorption by water vapour has been estimated by London [74], who used the absorption formula of Mügge and Möller [67], who in turn made use of Fowle's data. Recently, a

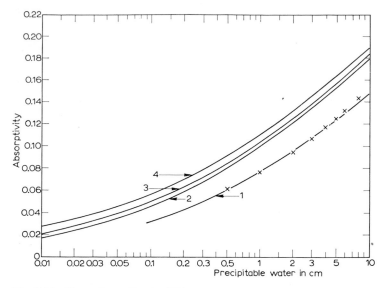

Fig. 3.32. Absorption estimates [65].
Curve 1: Absorptivities summed over the 0.72, 0.80, 0.94, 1.10, 1.38 and 1.87 μ H_2O bands.
       $\times$ = absorptivities estimated by McDonald [71] over the same bands.
Curve 2: Absorptivities over all the H_2O bands considered.
Curve 3: Absorptivities over the H_2O and CO_2 bands.
Curve 4: Absorptivities over the H_2O, CO_2 and O_2 bands.

similar estimate was reported by Roach [75] and by Yamamoto [65] who based their calculations mainly on the data of Howard *et al.* The former used the empirical absorption formulae proposed by Howard *et al.* and the latter made use of the absorption curves shown in Fig. 3.31. The absorption of O_2 bands was taken into account in Yamamoto's calculation.

Fig. 3.33 shows the energies absorbed by H_2O, CO_2 and O_2 bands as calculated by Yamamoto [65]. These calculations were based on the water vapour distribution for the mean July 10°–20°N atmosphere as compiled by London [76]. Because the H_2O and CO_2 bands overlap at about 2.7 μ, the corresponding absorptions cannot be accurately separated.

In Fig. 3.33 the CO_2 absorption represents the excess absorption by CO_2 over and above the H_2O absorption so that the minimum in the CO_2 absorption which occurs at about 400–500 mb is due to the rapid decrease in the excess contribution by the 2.7 μ CO_2 band in the lower troposphere. It will be seen from the figure that while the absorption by the H_2O bands predominates in the lower troposphere, the CO_2 absorption bands are the most important among the near-infrared bands in the stratosphere.

The heating rates for clear sky conditions which are due to the near-infrared absorption are shown in Fig. 3.34. The main features of the heating are as follows. The maximum heating occurs in the middle troposphere at about 4–6 km at all latitudes and for all seasons. The minimum heating occurs in the lower stratosphere, just above the tropopause. In the stratosphere the heating increases with height.

As long ago as 1912 Fowle [77] showed that the amount of water vapour in the vertical air column can be deduced from solar absorption measurements in the near-infrared water vapour bands. Fowle made use of the transmissivities in the ϱ (0.92 μ), Φ(1.1 μ) and ψ (1.4 μ) bands. The transmissivity at the centre of each band was measured as the ratio of the ordinate at the bottom of a band to the ordinate of the line drawn across the top of the band. Using the known relation between the transmissivity thus defined and the amount of water vapour, Fowle succeeded in evaluating the amount of water vapour in the atmosphere by measuring the solar transmissivity.

The solar absorption measurements have recently been used by Houghton et al. [78] and Gates et al. [79] to evaluate the amount of water vapour in the stratosphere with the aid of aeroplanes and balloons equipped with spectrometers. This seems to be a reliable method of determining the total amount of water vapour in the stratosphere. Their results are close to those obtained by Murgatroyd et al. [80] who made many aeroplane obser-

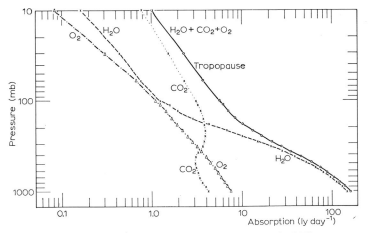

Fig. 3.33. Absorption of solar radiation at various altitudes [65].

vations using the frost-paint hygrometer originally developed by Brewer *et al.* [81]. However, the results of Houghton *et al.* [78] and Gates *et al.* [79] differ somewhat from those of Barret *et al.* [82] and Barclay *et al.* [83].

Until recently it was believed that the stratosphere is in an approximate radiative equilibrium. However, the effect of the general circulation and mixing in the stratosphere is needed to explain the ozone distribution and its variation with season and latitude. Detailed heat balance studies of the stratosphere are therefore becoming increasingly necessary.

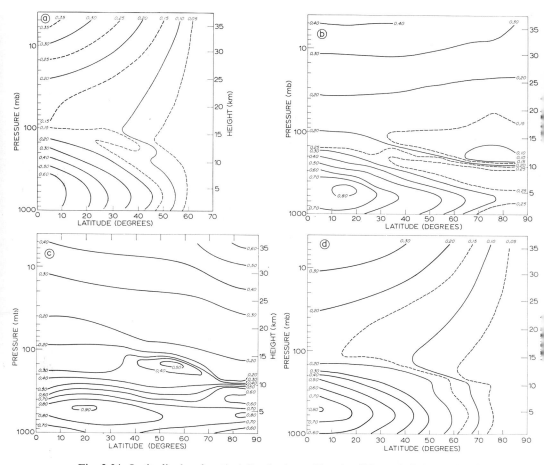

Fig. 3.34. Latitudinal and vertical distributions of heating (°C per day) due to the absorption by near-infrared bands [65].
(a) January, (b) April, (c) July, (d) October.

In the stratosphere the primary heat source is the ultraviolet solar energy absorbed by ozone, and the secondary source is the near-infrared solar radiation absorbed by water vapour and carbon dioxide. It is interesting to note that, according to Goody [84], the 9.6 μ ozone band is responsible for the heating of the stratosphere by absorbing terrestrial radiation arriving from the warm lower atmosphere. The radiative heating and cooling effects in the troposphere and lower stratosphere have been calculated by Manabe and Möller [85] whose results are shown in Fig. 3.35. The heating of the stratosphere is approximately compensated by both carbon dioxide (the 15 μ band) and water vapour (mainly the rotational band). The heating of the stratosphere due to the 9.6 μ ozone band is also noticed in Fig. 3.35.

As can be seen from Fig. 3.35 the radiative equilibrium is not established in the troposphere. The effect of radiative transfer, including the absorption of solar radiation, is overall cooling in the troposphere. Additional heat is supplied by turbulent transfer and by evaporation of water from the earth's surface into the atmosphere; it compensates the cooling due to radiation.

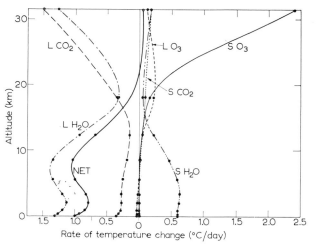

Fig. 3.35. Temperature changes (in °C per day for different altitudes) due to the absorption by various components [85].
S–O₃: due to the absorption of solar radiation by O_3.
S–H₂O: due to the absorption of solar radiation by H_2O.
S–CO₂: due to the absorption of solar radiation by CO_2.
L–O₃: due to the long-wave flux divergence by O_3.
L–H₂O: due to the long-wave flux divergence by H_2O.
L–CO₂: due to the long-wave flux divergence by CO_2.
Net: net radiative heating or cooling.

References pp. 107–110

7. ATMOSPHERIC TURBIDITY

Daily routine measurements of direct solar radiation are not made with spectral instruments but with pyrheliometers which are uniformly sensitive for the entire thermally effective spectrum $(0.3–3.0 \mu)$.

The intensity of direct solar radiation incident on a given surface element depends on the depletion along its path through the atmosphere. The depletion is small in pure air but increases with the amount of contamination, pollution, or turbidity associated with variable components such as water vapour, dust, haze, or generally, aerosol particles. Measurements of the direct solar radiation can therefore be used to determine the turbidity of the atmosphere.

There are, however, a number of difficulties. The depletion increases both with the extinction coefficient and the optical absolute air mass m. The latter is inversely proportional to the cosine of the zenith distance of the sum, or proportional to the relative air mass m_r, and directly proportional to the air pressure which is a measure of the air quantity in the vertical column above the observer (cf. Eq. (3.14), p. 52). The solar radiation is therefore found to vary with time of day, season, geographical latitude and altitude above sea level, even when the extinction coefficient remains constant.

In order to compare the contamination of the atmosphere at different times with the aid of measurements of direct solar radiation one must ensure that the latter are obtained for the same optical air mass, or approximately for the same elevation of the sun above the horizon. This primitive type of comparison, which is still occasionally used, fails when one tries to compare the turbidity at a high latitude, where the sun rises to small elevations above the horizon, with the turbidity in tropical areas, where the sun rapidly runs through the same elevations.

The mean extinction coefficient $a = m^{-1} \log (I_0/I)$ offers itself as a possible measure of the turbidity. With 'mean' an integral over all wavelengths is indicated.

The extinction coefficient can easily be computed from the above relation if the incident radiation intensity I_0 is known, the intensity I at the point under investigation is measured, and the absolute air mass $m = (p/1000)$ sec $z_\odot$ is determined from the air pressure p (in mb) and the zenith distance $z_\odot$ which is either measured or calculated from the apparent local time, the declination of the sun, and the geographical latitude.

A further difficulty is encountered in that the extinction coefficient is

TABLE 3.4. A daily series of measurements of direct solar radiation I, extinction coefficient a, and turbidity factor T (By courtesy of Dr. G. Dietze, Gotha, Germany).
Place: Gotha, 51°N, 11°E, 354 m; September 9, 1958.
$I_0 = 1.98$ cal $\cdot$ cm^{-2} $\cdot$ min^{-1}.

True local time	Air mass (m)	I (cal $\cdot$ cm^{-2} $\cdot$ min^{-1})	a	T
06.35	5.09	0.75	0.082	2.7
06.57	3.94	0.87	0.091	2.7
07.29	2.97	0.96	0.104	3.0
08.39	1.98	1.10	0.130	3.3
09.28	1.67	1.17	0.139	3.4
10.19	1.48	1.25	0.135	3.3
11.08	1.39	1.25	0.143	3.5
12.04	1.36	1.22	0.156	3.8
12.58	1.39	1.18	0.161	3.9
13.50	1.50	1.07	0.178	4.4
14.42	1.71	0.99	0.178	4.4
15.25	2.01	0.91	0.169	4.3
16.02	2.44	0.82	0.156	4.2
16.34	3.03	0.79	0.130	3.7
17.06	3.99	0.63	0.126	3.8
17.25	5.03	0.50	0.117	3.8

not even constant for constant turbidity when the optical air mass changes. This is clearly shown by measurements on clear days where a true change in the turbidity cannot be expected. Table 3.4 shows a series of measurements performed at the radiation observatory in Gotha, Germany.

The reason for these variations is the fact that the extinction coefficient is a function of the wavelength. This may be illustrated by the following schematic example. Suppose we divide the radiation incident on a depleting medium into two portions of equal intensity so that $I_0 = 2I_1 = 2I_2 \simeq 2$, and suppose further that the extinction coefficients are $a_1 = 0.1$ and $a_2 = 1.0$. We then have

$$I_c(m) = I_1 10^{-a_1 m} + I_2 10^{-a_2 m}$$

and since

$$I_c(m) = I_0 10^{-am}$$

we find that

$$a_{res} = a(m) = m^{-1} \log [I_0/I_c(m)]$$

The results of this simple calculation are given in Table 3.5.

TABLE 3.5. Schematic example of the variation of the extinction coefficient a with m, when I is composed of two regions with different extinctions: $a_1 = 0.1$, $a_2 = 1.0$.

m	I_1	I_2	$(I_1 + I_2)/2$	a_{res}
0	1.0	1.0	1.0	—
0.2	0.955	0.631	0.793	0.505
0.4	0.912	0.398	0.655	0.460
0.6	0.871	0.251	0.561	0.418
1.0	0.794	0.100	0.447	0.350
2	0.631	0.010	0.320	0.247
4	0.398	0.000	0.199	0.175
6	0.251	0.000	0.126	0.150

As can be seen from this table, the resultant extinction coefficient a_{res} is no longer constant but decreases with increasing m. For small m the extinction in region 1 is very small while the reverse situation occurs for large m. It follows that at first $a(m)$ is close to $a_2/2$ and then approaches a_1, thus decreasing with m. This decrease in the extinction coefficient a with air mass m is a general result. It is observed in direct solar radiation for which the depletion at short wavelengths is much larger than at long wavelengths. Because of its dependence on m, the mean extinction coefficient is not a suitable measure of atmospheric contamination.

In an atmosphere which contains neither vapour or other absorbers, nor haze or dust, the depletion of the direct solar radiation is described by the Rayleigh scattering coefficient

$$\sigma_\lambda = 0.00386 \, \lambda^{-4.05} \tag{3.26}$$

This is a simplified formula for the extinction by real anisotropic air molecules [86] (the value of Eq. (3.18) multiplied by log e). Several other proposals have been put forward with slightly different values for the coefficient and/or the exponent. Experimental data are as yet not sufficiently accurate to decide between these versions. Many tables, particularly those for the evaluation of the turbidity factor, are still based on the original Rayleigh formula for ideal molecules, in which the coefficient is 0.00356. Thus, the direct solar radiation is given by

$$I_a(m) = \int_0^\infty I_{0\lambda} 10^{-\sigma_\lambda m} d\lambda \tag{3.27}$$

and since by definition this must be equal to

$$I_a(m) = I_0 10^{-\sigma(m)m} \tag{3.28}$$

we obtain an equation defining the mean extinction coefficient $\sigma(m)$ of pure air for the total radiation of all wavelengths. This coefficient now proves to depend on m. Careful numerical integration yields the values of $I(m)$ and $\sigma(m)$ (Table 3.6).

Scattering by pure air is the basic atmospheric effect. Linke [89–91] therefore proposed to refer the mean extinction coefficient, observed in an actual turbid atmosphere, to the basic effect. Thus,

$$I(m) = I_0 \, 10^{-a(m)m} = I_0 \, 10^{-\sigma(m)mT} \tag{3.29}$$

where

$$T = [m\sigma(m)]^{-1} \log [I_0/I(m)] = P(m) \log [I_0/I(m)] \tag{3.30}$$

The quantity T was defined by Linke as the turbidity factor which indicates how many atmospheres of pure air produce the same total depletion (or give rise to the same intensity $I(m)$) of direct solar radiation, as measured by an instrument sensitive to all wavelengths.

Let us consider the turbidity factor in greater detail.

We shall assume that the turbidity of the atmosphere is due to water vapour and dust. Let w be the amount of water vapour in the vertical column above a horizontal unit area (measured in centimetres of precipitable water or in $g \cdot cm^{-2}$) and d the corresponding measure of dust in the vertical column. Suppose further that a_w and σ_d are the corresponding extinction coefficients. It was shown earlier that the wavelength dependence of σ_λ is $\sim \lambda^{-4}$ so that the scattering coefficient has its largest values in the ultraviolet and is very small in the infrared. Water vapour has no absorption in the visible and exhibits wide and strong absorption bands in the red and

TABLE 3.6. Extinction coefficient $\sigma(m)$, transmission factor $q(m)$, and intensity of the direct solar radiation $I(m)$ in an atmosphere of pure air. Extraterrestrial radiation $I_0 = 1.98 \, cal \cdot cm^{-2} \cdot min^{-1}$.

m	$\sigma(m)$	$q(m)$	$I(m)$ $(cal \cdot cm^{-2} \cdot min^{-1})$
0.5	0.0456	0.900	1.88
1.0	0.0431	0.906	1.79
1.5	0.0409	0.910	1.72
2	0.0388	0.915	1.66
3	0.0356	0.921	1.55
4	0.0331	0.927	1.46
6	0.0291	0.935	1.32
8	0.0262	0.941	1.22
10	0.0239	0.946	1.14

infrared. Scattering by haze particles is roughly proportional to $\lambda^{-\alpha}$, with α frequently between 1 and 2, and is therefore large in the violet and small in the red, but its variation with wavelength is slower than for air. These coefficients may be used to define a spectral or monochromatic turbidity factor T_λ which is given by

$$ma_\lambda T_\lambda = m\sigma_\lambda + m_r wa_{\lambda w} + m_r d\sigma_{\lambda d} \qquad (3.31)$$

Substituting $m = m_r p/p_0$ (Eq. (3.14)) we have the following expression for T:

$$T_\lambda = 1 + \frac{wa_{\lambda w} + d\sigma_{\lambda d}}{(p/p_0)\sigma_\lambda} \qquad (3.32)$$

On separating the haze component of the spectral turbidity factor it is found that

$$T_{\lambda d} - 1 = \frac{d\sigma_{\lambda d}}{(p/p_0)\sigma_\lambda} = \text{const. } \lambda^3 \qquad (3.33)$$

This part of the turbidity factor is small in the ultraviolet but large in the infrared. The water vapour component is given by

$$T_{\lambda w} - 1 = \frac{wa_{\lambda w}}{(p/p_0)\sigma_\lambda} \qquad (3.34)$$

The latter is very nearly zero in the visible but becomes exceptionally large in the region of the water vapour bands where $a_{\lambda w}$ is large and σ_λ small. It follows that the spectral turbidity factor T_λ is a complicated function of wavelength.

The overall turbidity factor corresponding to the entire spectrum is defined by the following expression

$$I_0 10^{-\sigma(m)mT} = \int_0^\infty I_{0\lambda} 10^{-\sigma_{\lambda a} m \left(1 + \frac{wa_{\lambda w}}{(p/p_0)\sigma_\lambda} + \frac{d\sigma_{\lambda d}}{(p/p_0)\sigma_\lambda}\right)} d\lambda \qquad (3.35)$$

Linke [89,90] was the first to suggest that the quantity T would be reasonably independent of the air mass or at least less dependent on m than the extinction coefficient for a turbid atmosphere. This has in fact proved to be correct and T has been employed as a useful measure of atmospheric turbidity.

In accordance with the definition of the turbidity factor given above, and in view of Eqs. (3.27) and (3.28), it is clear that T can never be less than unity. Thus it is easy to appreciate that a turbidity factor of 1.5 represents a rather clear atmosphere while a turbidity factor of say 5.0 refers to very turbid air. This straightforward relationship makes T a particularly useful parameter.

TABLE 3.7. Numerical values of the factor $P(m)$ as a function of the air mass m for computation of the turbidity factor T for total radiation.

m	$P(m)$	m	$P(m)$	m	$P(m)$	m	$P(m)$
0.5	43.9	2.0	12.86	4.0	7.55	6.0	5.72
0.6	37.0	2.2	11.91	4.2	7.30	6.4	5.48
0.7	32.1	2.4	11.11	4.4	7.06	6.8	5.28
0.8	28.4	2.6	10.43	4.6	6.84	7.2	5.09
0.9	25.5	2.8	9.85	4.8	6.64	7.6	4.92
1.0	23.2	3.0	9.33	5.0	6.46	8.0	4.77
1.2	19.8	3.2	8.90	5.2	6.29	8.4	4.63
1.4	17.3	3.4	8.51	5.4	6.13	8.8	4.51
1.6	15.5	3.6	8.15	5.6	5.98	9.2	4.39
1.8	14.0	3.8	7.84	5.8	5.85	9.6	4.29
2.0	12.9	4.0	7.55	6.0	5.72	10.0	4.19

The turbidity factor T can easily be computed with the aid of the quantity $P(m)$ given by Eq. (3.30). Numerical values of $P(m)$ are given in Table 3.7. In such calculations the measured intensity of solar radiation has to be reduced to the mean sun–earth distance by multiplying it by the factor $(R_d/R_0)^2$, where R_d is the sun–earth distance for the particular day, R_0 the annual mean. The logarithm of I_0/I is then computed, and the result multiplied by $P(m)$. A graphical computation chart is shown in Fig. 3.36.

In spite of the above advantages of the turbidity factor it suffers from one very important disadvantage. It is found that T is not strictly constant even when atmospheric turbidity does remain constant. As can be seen from Table 3.6, the mean extinction coefficient of pure air varies with the optical air mass. This variation is due to the dependence of σ_λ on wavelength $(\sim \lambda^{-4})$. From Eq. (3.35) it is evident that T can be independent of m only when the second and third term in brackets on the right side are constant and independent of λ. This is not the case, as is shown by Eqs. (3.33) and (3.34). The extinction by haze and water vapour differs from the extinction by pure air, and the deviation of the dependence on wavelength of both these extinctions gives rise to a diurnal variation in the turbidity factor even when w and d are constant. The variation is referred to as 'virtual variation' of the

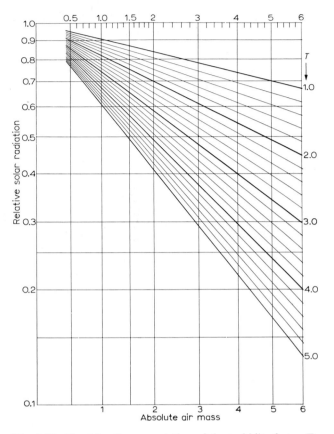

Fig. 3.36. Chart for the computation of the turbidity factor T.
Enter the diagram at the given value of the absolute air mass and at the given value of the relative solar radiation and read off the turbidity factor T from the inclined lines.

turbidity factor. The variation is different for different vapour or haze contents.

Hinzpeter [87,88] has reported some quantitative examples which show that T decreases between $m = 1$ and $m = 2$–3 and then increases again. Accordingly, one observes a noon maximum of T in the summer and also a decrease between early morning and forenoon, and an increase from afternoon towards sunset. Thus, some caution has to be exercised with regard to the observed diurnal variation of T.

Linke [91] recognised the difference in the spectral coefficients as the cause of the virtual variations. He therefore introduced the 'new turbidity factor' θ which is based on an atmosphere of pure air containing 1 cm of precipitable water. In this way he was able to calculate the new mean extinction coefficient $\varepsilon(m)$ with the aid of the relation

$$I_0 10^{-\varepsilon(m)m} = \int_0^\infty I_{0\lambda} 10^{-(\sigma_\lambda m + a_{\lambda w} m_r)} d\lambda \qquad (3.36)$$

This can be used to compute a quantity $\pi(m)$ which may be used to determine θ in the same way as $P(m)$ was used to determine T. However, this new parameter is unfortunately still subject to virtual variations although these variations are smaller than in the case of T. The new turbidity factor can assume values smaller than unity when the water vapour content is very small or at least smaller than 1 cm. The new turbidity factor is not as easy to interpret as T. Diagrams which may be used in computing θ have been given by Danc [92].

In the evaluation of radiation measurements at high altitudes the local absolute air mass must be used in the calculation of the turbidity factor (cf. Eq. (3.14)). When the atmosphere above the high altitude station contains the same amount of 'contamination', e.g. water vapour, as the atmosphere of a sea-level station, the turbidity factor ought to be the same in both cases. However, it is possible to show by a consideration of synthetic cases that the turbidity factor would in fact be different. It is therefore necessary to have a conversion formula so that high altitude results can be reduced to sea level. This can be done with the aid of the formula given by Feussner and Dubois [99]

$$T_p = 1 + (T-1) \frac{P(m_r)}{P(m)} \qquad (3.37)$$

where T_p is the extrapolated turbidity factor. This type of calculation should be used when the air pressure at the station differs from 1000 mb by more than 50 mb. The necessary numerical values of $P(m)$ may be taken from Table 3.7.

Water vapour and dust (haze) are the most important sources of turbidity. They can be separated out because of their widely different contributions in the visible and infrared parts of the spectrum. The separation can easily be achieved by means of the Schott red filter RG-2, whose transmission has a sharp cut-off at 630 mμ. If $DR2$ is the ratio of the radiation with $\lambda > 630$ mμ incident at the filter, to the transmitted radiation I_R which is measured, then the quantity given by

$$I_k = I_t - DR2 \cdot I_R$$

is approximately the sum of the visible and ultraviolet portions of the solar radiation which is frequently (though unfortunately) referred to as the shortwave radiation ('Kurzstrahlung' in Linke's terminology). It is the total radiation measured without the filter. I_k can then be used to determine the Kurzstrahlung turbidity factor which is a good measure of the haze or aerosol content of the atmosphere. Similarly, the turbidity factor based on the red component I_R can be used as a measure of the water content of the atmosphere.

A very extensive survey of measurements of turbidity factors all over the world has been published by Steinhauser [93]. It turns out that in the tropics T is much larger than in the clean polar air which is poor in water vapour. Unfortunately it is not always possible to compare average values obtained at different places because of the different procedures adopted by different observers, some of whom measure direct solar radiation on clear and cloudless days only, while others make use of each gap in the clouds to make a quick measurement. A survey of different synoptic air masses measured at the same place is often more reliable (Table 3.8).

Similar tables have been reported for the USA, Japan, Moscow and many other places. It is found that there are considerable differences between large cities and the open country. A most impressive example has been reported for Vienna. This even shows that the largest turbidity is mostly found at the lee side of the city.

The change in the turbidity factor with altitude is particularly instructive. Many measurements of this type are available for the Austrian Alps. Steinhauser [93] has reviewed the annual variation in the central European climate as well as the decrease in the turbidity with altitude (Table 3.9).

TABLE 3.8. Turbidity factors in various synoptic air masses (averaged values) after measurements in Potsdam, Germany [88].

Air mass	T
Continental arctic air	2.20
Maritime arctic air	2.10
Continental moderate air	2.89
Maritime moderate air	2.85
Continental tropical air	3.76
Maritime tropical air	3.66

TABLE 3.9. Annual variation of the turbidity factor T in the Austrian alps [93].

Month	Altitude (metres)				
	200	500	1000	2000	3000
January	3.0	2.7	2.2	1.8	1.7
February	2.8	2.5	2.1	1.8	1.7
March	3.2	2.9	2.6	2.1	1.9
April	3.6	3.3	2.8	2.2	2.1
May	3.8	3.6	3.2	2.5	2.2
June	3.8	3.5	3.2	2.6	2.2
July	3.9	3.7	3.3	2.6	2.2
August	3.8	3.4	3.1	2.6	2.2
September	3.3	3.0	2.8	2.3	2.0
October	3.2	2.8	2.5	2.0	1.9
November	2.7	2.4	2.2	1.8	1.8
December	2.6	2.4	2.1	1.7	1.7

However, the stratification of the atmosphere is more clearly exhibited by the 'special turbidity factor' T^* given by Linke [90]. This quantity specifies the turbidity in an air layer between two altitudes and is given by

$$I_l = I_u 10^{-T^*[m_l\sigma(m_l)-m_u\sigma(m_u)]} \tag{3.38}$$

or more explicitly by

$$T^* = \frac{\log I_u - \log I_l}{1/P(m_l) - 1/P(m_u)} \tag{3.39}$$

where u is the upper height, l is the lower height and I_u and I_l are the corresponding radiation intensities. An analogous equation may be written for the special Kurzstrahlung turbidity factor T_k^*.

Steinhauser's determinations of the special turbidity factor are given in Table 3.10. These figures refer to the total radiation of all wavelengths and

TABLE 3.10. Special turbidity factor T^* of different altitude layers in the Austrian alps [93].

Season	Layer (km)			
	0.2–1.0	1.0–1.5	1.5–2.0	2.0–3.0
Winter	11.0	6.1	4.7	2.2
Spring	10.7	10.3	6.8	4.0
Summer	11.1	9.5	7.6	5.7
Autumn	9.6	8.2	5.4	3.0

therefore contain dust and water vapour effects. The dust component may be extracted by means of the special Kurzstrahlung turbidity factor T_k^*. Table 3.11 shows Linke's evaluation of balloon measurements on March 4, 1938 (high pressure situation), indicating the considerable increase in the aerosol concentration near the surface.

TABLE 3.11. Special turbidity factor T_k^* during a balloon flight, March 4, 1938, from Frankfurt/M., Germany (high pressure situation) [91].

Height (km)	0.9	0.55	1.27	1.66	2.30
T_k^*		2.9	4.6	3.8	2.3

Penndorf has evaluated a series of eight ascents by Krug. These measurements also indicate the very steep decrease of the haze turbidity with altitude and the very large amounts near the surface (Table 3.12). High-altitude data of this type are unfortunately very rare.

TABLE 3.12. Mean special turbidity factor T_k^* during 8 flights [100,101].

Height (km)	0.57	1	2	3	4	5	6	7	8	9
T_k^*		7.4	3.7	2.1	1.7	1.3	1.4	1.4	1.8	1.4

The extinction coefficient $\sigma_{\lambda d}$ of haze follows a law $\sigma_{\lambda d} \sim \lambda^{-\alpha}$ when the size spectrum of aerosol particles is of the form

$$dn = cr^{-(\alpha+3)}dr \tag{3.40}$$

where r is the particle radius, n is the number of particles per unit volume per unit radius interval and c is a constant.

The above particle-size distribution is a mean distribution and considerable deviations from it may occur in isolated cases. In effect, the exponent α in the above expression for the extinction coefficient may vary between 0.5 and 3.0. Ångström [94] used early measurements of the spectral extinction and found that

$$\sigma_{\lambda d} \sim \lambda^{-1.3}$$

Later measurements confirmed this value of the exponent for some industrial regions. In the Kurzstrahlung range (*i.e.* in the range corresponding to the difference between the total radiation and the radiation transmitted by the RG-2 red filter), the extinction is almost entirely due to pure air and haze. For this spectral range Ångström suggested the relation

$$I_k = \int_0^{0.630\ \mu} I_0 10^{-(m_0\lambda + m_r\beta\lambda^{-1.3})} \mathrm{d}\lambda \qquad (3.41)$$

and referred to β as the turbidity coefficient. Comparison with Eq. (3.35) shows that β is a measure of the haze content of the atmosphere.

By confining one's attention to the shortwave part of the solar spectrum (Kurzstrahlung) in the definition of turbidity coefficient, the disturbances due to selective absorption by water vapour are avoided. A simple evaluation based on this relation cannot however be used to compute β.

Fig. 3.37 is based on tables of I_k as a function of m and β reported by Hoelper [97] and Feussner [98]. In order to use this chart, the measured shortwave radiation is reduced to the mean sun–earth distance by means of the relation $I_k = I_{k,\mathrm{meas}}(R_0/R_d)^2$ and the value of β corresponding to given m, I_k can easily be read off.

Fig. 3.37 refers to measurements at stations at which the pressure is approximately 1000 mb. For pressures which are very different from this value analogous calculations yield unrealistic values of β. They may, however, be reduced to the mean sea-level values by means of the simple formula

$$\beta = \beta_p(p/1000) \qquad (3.42)$$

As mentioned above, the exponent in Eq. (3.40) undergoes variations due to changes in the size distribution of the particles. This gives rise to errors in β which appear as daily virtual variations in the turbidity coefficients.

It therefore appears that the only way to avoid virtual variations in measures of turbidity is to determine the turbidity factor or the turbidity coefficient together with an independent simultaneous determination of the wavelength exponent. Schüepp [95] has suggested the relation

$$d\sigma_{\lambda d} = B^{-\alpha}$$

for the extinction due to atmosphere aerosols where B may be called the turbidity coefficient. In order to determine B and α it is necessary to have two measured quantities. This can be achieved with the aid of the orange Schott filter OG-1 which has a sharp cut-off at 525 mμ, similar to the 630 mμ cut-off for the red filter.

If $DR1$ is the ratio of the incident radiation for $\lambda > 525$ mμ and the

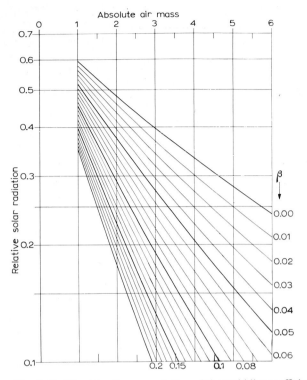

Fig. 3.37. Chart for the computation of the turbidity coefficient β.
Enter the diagram at the given value of the absolute air mass and at the given value of the relative solar radiation, measured with the red filter RG-2, and read off the turbidity coefficient β from the inclined lines.

transmitted radiation I_G which is measured, then, as in the case of the red filter, the blue radiation intensity is given by

$$I_b = I_t - DR1 \cdot I_G$$

Moreover, yellow radiation I_y may be defined by

$$I_y = DR1 \cdot I_G - DR2 \cdot I_R$$

and may be used in the determination of B and α in addition to I_b.

Graphical procedures have been devised by Schüepp [95] and have recently been adopted for current usage by Valko [96]. It appears, however, that the method is very sensitive to experimental errors and extreme care is necessary in calibrating the instruments and in cleaning the filters. Averages of several measurements have to be used to determine α, since single measurements are not sufficient reliable. Daily mean values of α may then be used to evaluate B, the red and infrared radiation I_R, and the precipitable water w in the atmosphere to an accuracy comparable with the accuracy of aerological measurements whenever turbidity is small. When the turbidity is high, actinometric measurements can no longer be used to determine useful w values. The method is ingenious but brings us to the limits of actinometric measurements.

Linke's turbidity factor T is the only measure of turbidity which is suitable for an evaluation of total-radiation measurements. It is an indicator of the amount of haze and water vapour in the air. This is also valid for the new turbidity factor $θ$ which has smaller virtual variations.

Ångström's turbidity coefficient and the Kurzstrahlung turbidity factor are defined for the spectral region transmitted by the Schott red filter RG-2 and they are therefore measures of the haze content only. The turbidity coefficient has the advantage of a better physical definition.

Schüepp's generalized turbidity coefficient B, taken together with the wavelength exponent α and the precipitable water content w, gives the most valuable information because the size spectrum of the aerosol particles can be inferred from α. Thus the method provides more information about the nature of the aerosols and the atmospheric humidity than T and $β$.

More detailed analyses of the spectral turbidity may be made with the aid of interference filters or spectrographs. Such instruments, however, require frequent calibration and tests, making the measurements a difficult experimental task.

REFERENCES

1. F. LINKE, Die Sonnenstrahlung und ihre Schwächung in der Atmosphäre, in *Handbuch der Geophysik*, Bornträger, Berlin, Vol. 8 (1942), pp. 239–332.
2. H. C. VAN DE HULST, *Light Scattering by Small Particles*, Wiley, New York (1957).
3. R. PENNDORF, Tables of Refractive Index for Standard Air and the Rayleigh Scattering Coefficient for the Spectral Region between 0.2 and 20.0 $μ$ and their Application to Atmospheric Optics, *Geophys. Res. Directorate (U.S.), Air Force Cambridge Res. Center*, Project No. 7621, April 1955 *(AFCRC–TN–55–206)*.
4. H. GAERTNER, The Transmission of Infrared in Cloudy Atmosphere, *Navord Report*, No. 429 (1947).
5. K. L. COULSON, *Planetary Space Sci.*, 1 (1959) 270–284.
6. R. PENNDORF, Results of an Approximation Method to the Mie Theory for Colloidal Spheres, Sci. Rept. No. 1, *AFCRC–TN–59–608, Contract No. AF 19(604)–5743* (1959).

7. S. FRITZ, *Sci. Monthly*, 84 (1957) 55–65.
8. R. S. FRASER, Scattering of Atmosphere Aerosols, *Dept. of Meteorol. (U.S.), UCLA, Sci. Rept.*, No. 2, *Contract No. AF 19(604)–2429*, October 1959.
9. S. FRITZ, *J. Opt. Soc. Am.*, 45 (1955) 820–825.
10. C. DORNO, *Veröffentl. Preuss. Meteorol. Inst., Abhandl.*, 8, No. 303 (1919).
11. B. V. VEYNBERG, *Natural Illumination of Schools* (in Russian), Moscow (1951).
12. F. TONNE AND W. NORMANN, *Z. Meteorol.*, 14 (1960) 166–179.
13. E. R. PYASKOVSKAYA-FESENKOVA, *Investigation of the Scattered Light in the Earth's Atmosphere* (in Russian), Moscow (1957), p. 182.
14. R. A. CRAIG, *Meteorol. Monographs*, 1, No. 2 (1950).
15. K. WATANABE, M. ZELIKOFF AND C. Y. INN, *AFCRC Tech. Rept.*, No. 52–53, *Geophys. Res. Papers (U.S.)* No. 21 (1953).
16. K. WATANABE AND F. F. MARMO, *J. Chem. Phys.*, 25 (1956) 965.
17. P. LEE, *J. Opt. Soc. Am.*, 45 (1955) 703.
18. K. WATANABE, *Advan. Geophys.*, 6 (1958) 187.
19. R. W. DITCHBURN, J. E. S. BRADLEY, C. G. CANNON AND G. MUNDAY, *Rocket Exploration of the Upper Atmosphere*, Pergamon, London (1954), pp. 327–334.
20. D. R. BATES AND M. J. SEATON, *Proc. Phys. Soc. (London)*, B 63 (1950) 129.
21. M. NICOLET, *1952 Yearbook of the University of New Mexico*, Univ. of New Mexico Press, Albuquerque, N.M. (1952), Chapter 12.
22. T. Z. NY AND S. P. CHOONG, *Compt. Rend.*, 195 (1933) 309.
23. E. VIGROUX, *Ann. Phys. (Paris)*, [12] 8 (1953) 709.
24. E. C. Y. INN AND Y. TANAKA, *J. Opt. Soc. Am.*, 43 (1953) 870.
25. Y. TANAKA, E. C. Y. INN AND K. WATANABE, *J. Chem. Phys.*, 21 (1953) 1654.
26. H. FRIEDMAN, *Physics of the Upper Atmosphere* (edited by J. R. Ratcliffe), Academic Press, New York (1960), p. 134.
27. S. CHAPMAN, *Proc. Phys. Soc. (London)*, 43 (1931) 26.
28. M. NICOLET, *Inst. Roy. Meteorol. Belg.*, *Mem.* No. 19 (1945).
29. H. FRIEDMAN, *Physics of the Upper Atmosphere* (edited by J. R. Ratcliffe), Academic Press, New York (1960), p. 197.
30. O. R. WULF AND L. S. DEMING, *Terrest. Magnetism Atmospheric Elec.*, 43 (1938) 283.
31. E. O. HULBERT, *Phys. Rev.*, 53 (1938) 344.
32. L. VEGARD, *Geofys. Publikasjoner*, 12, No. 5 (1938) 1–23.
33. F. HOYLE AND D. R. BATES, *Terrest. Magnetism Atmospheric Elec.*, 53 (1948) 51.
34. E. T. BYRAM, T. A. CHUBB AND H. FRIEDMAN, *Phys. Rev.*, 92 (1953) 1066; *J. Geophys. Res.*, 61 (1956) 251.
35. S. CHAPMAN, *Mem. Roy. Meteorol. Soc.*, 3 (1930) 103.
36. D. R. BATES AND M. NICOLET, *J. Geophys. Res.*, 55 (1950) 301.
37. F. S. JOHNSON, J. D. PURCELL, R. TOUSEY AND K. WATANABE, *J. Geophys. Res.*, 57 (1952) 157.
38. K. WATANABE AND M. ZELIKOFF, *Advan. Geophys.*, 6 (1958) 206.
39. D. R. BATES, The Physics of the Upper Atmosphere, in *The Earth as a Planet* (edited by G. P. Kuiper), Univ. of Chicago Press, Chicago, Ill. (1954), p. 583.
40. G. M. B. DOBSON, *Proc. Roy. Soc. (London)*, A 129 (1930) 411.
41. G. M. B. DOBSON, D. N. HARRISON AND J. LAWRENCE, *Proc. Roy. Soc. (London)*, A 122 (1929) 456.
42. E. TØNSBERG AND K. L. OLSEN, *Geofys. Publikasjoner*, 13, No. 12 (1944) 1–39.
43. O. R. WULF AND L. S. DEMING, *Terrest. Magnetism Atmospheric Elec.*, 42 (1937) 195.
44. J. PRESSMAN, *AFCRC–TR–54–202*, *Geophys. Res. Papers (U.S.)*, No. 33 (1954).
45. G. H. KIEKE AND H. D. BABCOCK, *Proc. Natl. Acad. Sci. U.S.*, 13 (1927) 670.
46. W. F. GIAUQUE AND H. L. JOHNSTON, *J. Am. Chem. Soc.*, 51 (1929) 1436.

47. L. GOLDBERG, The Absorption Spectrum of the Atmosphere, in *The Earth as a Planet* (edited by G. P. Kuiper), Univ. of Chicago Press, Chicago, Ill. (1954).
48. H. D. BABCOCK AND C. E. MOORE, *The Solar Spectrum, 6600 to 13495 Å*, Carnegie Institution of Washington, Washington, D.C. (1947).
49. O. C. MOHLER, A. K. PIERCE, R. R. MCMATH AND L. GOLDBERG, *Photometric Atlas of the Near Infrared Solar Spectrum*, Univ. of Michigan Press, Ann Arbor, Mich. (1951).
50. J. H. SHAW, R. M. CHAPMAN, J. N. HOWARD AND M. L. OXHOLM, *Astrophys. J.*, 113 (1951) 268.
51. J. H. SHAW, M. L. OXHOLM AND H. H. CLAASEN, *Ohio State Univ., Sci. Rept.*, No. 1A–4; *Contract No. AF 19(122)65* (1951).
52. M. MIGEOTTE, L. NEVEN AND J. SWENSSON, The solar spectrum from 2.8 to 23.7 μ, *Mem. Soc. Roy. Sci. Liege*, Special volume (1957).
53. J. N. HOWARD, D. E. BURCH AND D. WILLIAMS, *AFCRC–TR–55–213, Geophys. Res. Papers (U.S.)*, No. 40 (1955).
54. G. N. PLASS AND D. I. FIVEL, *Astrophys. J.*, 117 (1953) 225.
55. R. LADENBURG AND F. REICHE, *Ann. Physik*, 42 (1913) 181.
56. W. M. ELSASSER, Heat Transfer by Infrared Radiation, *Harvard Meteorol. Studies*, No. 6, Harvard Univ. Press, Cambridge, Mass. (1942).
57. R. M. GOODY, *Quart. J. Roy. Meteorol. Soc.*, 78 (1952) 165.
58. L. D. KAPLAN, *Proc. Toronto Meteorol. Conf.*, (1953) 43.
59. G. N. PLASS, *J. Opt. Soc. Am.*, 48 (1958) 690; 50 (1960) 868.
60. F. E. FOWLE, *Smithsonian Inst. Misc. Collections*, 68, No. 8 (1917) 1.
61. V. R. STULL, P. J. WYATT AND G. N. PLASS, *AF 04(695)96* (1963).
62. D. E. BURCH, D. GRYVNAK AND D. WILLIAMS, *Geophys. Res. Directorate (U.S.), AFCRL, Sci. Rept.*, No. 2, *Contract No. AF 19(604)–2633* (1960).
63. N. S. BENEDICT, *Contract No. AF 19(604)–1001* (1956).
64. H. J. KOSTKOWSKI, *Office of Naval Research (U.S.), Contract No. 248(01)* (1955).
65. G. YAMAMOTO, *J. Atmospheric Sci.*, 19 (1962) 182.
66. H. H. KIMBALL, *Monthly Weather Rev.*, 58 (1930) 43–52.
67. R. MÜGGE AND F. MÖLLER, *Z. Geophys.*, 8 (1932) 53–64.
68. H. J. TANCK, *Ann. Hydrogr. mar. Meteorol.*, 6 (1940) 47–64.
69. G. YAMAMOTO AND G. ONISHI, *J. Meteorol.*, 9 (1952) 415–421.
70. H. G. HOUGHTON, *J. Meteorol.*, 11 (1954) 1–9.
71. J. E. MCDONALD, *J. Meteorol.*, 17 (1960) 319–328.
72. S. P. LANGLEY AND C. G. ABBOT, *Ann. Astrophys. Obs. Smithsonian Inst.*, 1 (1900).
73. D. M. GATES, *J. Opt. Soc. Am.*, 50 (1960) 1299–1304.
74. J. LONDON, *J. Meteorol.*, 9 (1952) 145–151.
75. W. J. ROACH, *Quart. J. Roy. Meteorol. Soc.*, 87 (1961) 364–373.
76. J. LONDON, *AFCRC–TR–57–287* (1957).
77. F. E. FOWLE, *Astrophys. J.*, 35 (1912) 140.
78. J. T. HOUGHTON AND J. S. SEELEY, *Quart. J. Roy. Meteorol. Soc.*, 86 (1960) 358.
79. D. M. GATES, D. C. MURCRAY AND C. C. SHAW, *J. Opt. Soc. Am.*, 48 (1958) 1010.
80. R. J. MURGATROYD, P. GOLDSMITH AND W. E. H. HOLLINGS, *Quart. J. Roy. Meteorol. Soc.*, 81 (1955) 533.
81. A. W. BREWER, B. CWILONG AND C. M. B. DOBSON, *Proc. Phys. Soc. (London)*, 60 (1948) 52.
82. E. W. BARRET, L. R. HERNDON AND H. J. CARTER, *Tellus*, 2 (1950) 302.
83. F. R. BARCLAY, M. J. W. ELLIOTT, P. GOLDSMITH AND J. V. JELLEY, *Quart. J. Roy. Meteorol. Soc.*, 86 (1960) 259.
84. R. M. GOODY, *Proc. Roy. Soc. (London)*, A 197 (1949) 487.
85. S. MANABE AND F. MÖLLER, *Monthly Weather Rev.*, 89 (1961) 503.

86. J. CABANNES, *Thesis*, Paris (1921).
87. H. HINZPETER, *Z. Meteorol.*, 4 (1950) 1.
88. L. FOITZIK AND H. HINZPETER, *Sonnenstrahlung und Lufttrübung* (Probleme der kosmischen Physik, Band 31), Akad. Verlagsges., Leipzig (1958).
89. F. LINKE AND K. BODA, *Meteorol. Z.*, 39 (1922) 161.
90. F. LINKE, *Beitr. Phys. Freien Atmosphäre*, 15 (1929) 176.
91. F. LINKE, Die Sonnenstrahlung und ihre Schwächung in der Atmosphäre, in *Handbuch der Geophysik*, Bornträger, Berlin, Vol. 8 (1942), p. 281.
92. J. DANC, *Meteorol. Zpravy*, 8 (1955) 71.
93. F. STEINHAUSER, *Meteorol. Z.*, 56 (1939) 172.
94. A. ÅNGSTRÖM, *Geografiska Annaler*, 11 (1929) 156; 12 (1930) 130.
95. W. SCHÜEPP, *Arch. Meteorol. Geophys. Bioklimatol. Ser. B*, 1 (1949) 257.
96. P. VALKO, *Arch. Meteorol. Geophys. Bioklimatol. Ser. B*, 11 (1962) 75–107.
97. O. HOELPER, *Wiss. Abhandl. Reichsamt Wetterdienst*, 5, No. 10 (1939).
98. K. FEUSSNER, *Ber. Preuss. Meteorol. Inst.*, 1931 (1932) 89.
99. K. FEUSSNER AND P. DUBOIS, *Gerlands Beitr. Geophys.*, 27 (1930) 132.
100. U. KRUG-PIELSTICKER, *Ber. Deut. Wetterdienstes U.S. Zone*, No. 8 (1949).
101. R. PENNDORF, *Geophys. Res. Papers (U.S.)*, No. 25 (1954).
102. K. YA. KONDRATYEV, *Radiant Solar Energy* (in Russian), Leningrad (1954).
103. F. MÖLLER, Strahlung in der unteren Atmosphäre, in *Handbuch der Physik (Encyclopaedia of Physics)*, Springer, Berlin, Vol. 48 (1957), pp. 155–253.
104. IGY Instruction Manual, Part VI: Radiation Instruments and Measurements, *Annals of IGY*, V (1957), Pergamon, London.
105. F. VOLZ, *Ber. Deut. Wetterdienstes*, 2, No. 13 (1954).
106. F. VOLZ, Optik des Dunstes, in *Handbuch der Geophysik*, Bornträger, Berlin, Vol. 8 (1942), pp. 822–894.

Chapter 4

DIRECT AND SCATTERED RADIATION REACHING THE EARTH, AS INFLUENCED BY ATMOSPHERIC, GEOGRAPHICAL AND ASTRONOMICAL FACTORS

1. INTRODUCTION

There are two basic ways in which the geographical distribution of solar radiation may be studied. One of them involves measurements by a network of closely spaced stations, and the other is based on use of physical formulae and constants. In practice a combination of both methods must be used, in order to achieve an accuracy which is sufficient for most applications. Many of the physical constants can only be evaluated at a very small number of specialized institutions, while measurements of radiation falling on surfaces other than the horizontal plane, or the plane normal to the solar beam, are virtually non-existent for regions where they are needed most.

In this chapter a theoretical treatment of the problem will be given, together with a number of tables, graphs and formulae. Data obtained at different observatories are reviewed, and whenever possible a comparison is made between theory and observation. It must, however, be emphasized that some of the physical coefficients are not known to a sufficient degree of accuracy, and much research in this direction is being done, using the data published by the IGY Data Centre.

The intensity falling on a plane receiver is influenced to a greater or lesser extent by a large number of effects which may be classified as astronomical, geographical, geometrical, physical and meteorological. These effects are listed below.

Astronomical factors

(1) The solar spectrum between 0.30 μ and 5.0 μ and the magnitude of the solar constant I_0.

(2) Variation with the earth–sun distance S.

(3) Variation with the solar declination δ.

(4) Variation with the time angle t.

Geographical factors

(5) Variation with the latitude φ of the station.

(6) Variation with the longitude Λ of the station.

(7) Dependence on the height h above sea level (preferably expressed in terms of the mean pressure p).

Geometrical factors

(8) Dependence on the altitude γ of the sun.

(9) Dependence on the azimuth $A_\odot$ of the sun.

(10) The effect of the tilt of the receiving plane, ε, relative to the horizon.

(11) Variation with the azimuth of the tilted plane A_p.

Physical factors

(12) Extinction by pure atmosphere, a (Rayleigh extinction).

(13) Water content of the atmosphere, w (expressed in centimetres of precipitable water).

(14) The turbidity coefficient B of Ångström-Schüepp.

(15) The exponent α in Ångström's turbidity formula.

(16) The ozone content of the atmosphere (expressed in cm NTP, IGY scale 1957).

Meteorological factors

(17) The effect of the cloudiness of the sky, C.

(18) The effect of the albedo of the ground, A.

Only three of these elements can be completely evaluated from other available data, namely, the effects of the longitude Λ, the altitude γ of the sun, and the azimuth $A_\odot$ of the sun. Measurements carried out in different climates appear to show that in statistical calculations the exponent α in

Ångström's formula may be taken as 1.5, so long as only total radiation is considered. Space and time variations in the ozone concentration have little effect on the global radiation, so that the mean value of 0.34 cm (new IGY scale) may be used. We are thus left with thirteen other effects.

The appropriate formulae for the angle of elevation γ ($=90° - z_\odot$, where $z_\odot =$ zenith distance) and the azimuth angle $A_\odot$ of the sun as functions of the geographical latitude φ, longitude Λ, the time angle t and the declination δ have already been given in Chapter 2. A graphical method is described in the appendix to this chapter (see Diagram 4.1, p. 148). The relation between ε, A_p and the angle of incidence β is given by Eq. (2.6) (p. 41). Finally, the solar spectrum outside the atmosphere and the reduction of the mean distance from the sun must be considered (cf. Chapter 2).

2. CLEAR SKY CONDITIONS

According to Ångström [1] the extinction due to the atmosphere depends on three physical processes which affect the various regions of the solar spectrum in a rather different way. The three effects are represented by the Rayleigh extinction coefficient a_A (scattering by pure atmosphere), the extinction coefficient a_d which represents scattering by dust (aerosols), and the extinction coefficient a_w which represents absorption, mainly by water vapour.

The complete Ångström turbidity formula giving the direct component I_γ of solar radiation on a plane perpendicular to the solar beam is

$$I_\gamma = (R_0/R)^2 \int_0^\infty I_{0,\lambda} 10^{-(ma_A + m_r a_d + a_w)} \, d\lambda \qquad (4.1)$$

where $(R_0/R)^2$ is the mean-distance correction factor (cf. Chapter 2), m_r is the Bemporad function (relative air mass) which is unity when the sun is in the zenith (Table 3.1, p. 51), $m = pm_r/1000$ is the air mass for stations at which the pressure is p mb, $a_A = 0.00386 \, \lambda^{-4.05}$ is the Rayleigh extinction coefficient [2] including corrections for refraction and anisotropy of the air molecules [3], $a_d = B(2\lambda)^{-\alpha}$ is the coefficient representing extinction by dust (we shall assume that $\alpha = 1.5$), and a_w represents absorption by water vapour, and the less important absorption by ozone, carbon dioxide and other gases.

No explicit formulae are available for the absorption. We shall use the results of Fowle [4] and of Mörikofer and Schüepp [5]. Table 4.1 gives

TABLE 4.1. Values of a_w for absorption by water vapour, carbon dioxide and oxygen.

$m_r \cdot w$ (cm)	Values of a_w for various wavelengths						
	0.72–0.80 μ	0.80–1.0 μ	1.0–1.25 μ	1.25–1.5 μ	1.5–2.0 μ	2.0–3.0 μ	>3.0 μ
0.10	0.001	0.008	0.017	0.097	0.066	0.190	0.233
0.20	0.002	0.012	0.023	0.143	0.092	0.256	0.312
0.40	0.004	0.019	0.033	0.203	0.120	0.328	0.403
0.70	0.007	0.027	0.047	0.260	0.144	0.402	0.492
1.00	0.011	0.035	0.058	0.302	0.160	0.435	0.562
2.00	0.022	0.053	0.087	0.388	0.191	0.566	0.715
4.00	0.040	0.082	0.129	0.483	0.223	0.683	0.907
7.00	0.067	0.118	0.177	0.567	0.249	0.786	1.082
10.00	0.081	0.150	0.210	0.620	0.266	0.863	1.205
20.00	(0.138)	(0.202)	(0.272)	(0.720)	(0.295)	(1.000)	(1.435)
30.00	(0.165)	(0.240)	(0.315)	(0.780)	(0.315)	(1.085)	(1.570)

the values of a_w for water vapour, carbon dioxide and oxygen. In order to be able to use these data, the geographical and seasonal variations of B, w and the ozone content must be known.

Ozone exerts a rather slight effect, and the mean seasonal values ob-

TABLE 4.2. Seasonal variation of atmospheric ozone in cm NTP (new IGY scale) [6,7].

Lati-tude	Month											
	Jan.	Feb.	Mar.	Apr.	May	June	July	Aug.	Sept.	Oct.	Nov.	Dec.
90°N	0.33	0.39	0.46	0.42	0.39	0.34	0.32	0.30	0.27	0.26	0.28	0.30
80°N	0.34	0.40	0.46	0.43	0.40	0.36	0.33	0.30	0.28	0.27	0.29	0.31
70°N	0.34	0.40	0.45	0.42	0.40	0.36	0.34	0.31	0.29	0.28	0.29	0.31
60°N	0.33	0.39	0.42	0.40	0.39	0.36	0.34	0.32	0.30	0.28	0.30	0.31
50°N	0.32	0.36	0.38	0.38	0.37	0.35	0.33	0.31	0.30	0.28	0.29	0.30
40°N	0.30	0.32	0.33	0.34	0.34	0.33	0.31	0.30	0.28	0.27	0.28	0.29
30°N	0.27	0.28	0.29	0.30	0.30	0.30	0.29	0.28	0.27	0.26	0.26	0.27
20°N	0.24	0.26	0.26	0.27	0.28	0.27	0.26	0.26	0.26	0.25	0.25	0.25
10°N	0.23	0.24	0.24	0.25	0.26	0.25	0.25	0.24	0.24	0.23	0.23	0.23
0°	0.22	0.22	0.23	0.23	0.24	0.24	0.24	0.23	0.23	0.22	0.22	0.22
10°S	0.23	0.24	0.24	0.24	0.24	0.24	0.24	0.24	0.24	0.24	0.24	0.23
20°S	0.24	0.25	0.24	0.25	0.25	0.25	0.25	0.26	0.26	0.26	0.26	0.25
30°S	0.27	0.28	0.26	0.27	0.28	0.28	0.29	0.31	0.32	0.32	0.29	0.29
40°S	0.30	0.29	0.28	0.29	0.31	0.33	0.35	0.37	0.38	0.37	0.34	0.32
50°S	0.31	0.30	0.29	0.30	0.32	0.36	0.39	0.40	0.40	0.39	0.37	0.35
60°S	0.32	0.31	0.30	0.30	0.33	0.38	0.41	0.42	0.42	0.40	0.39	0.35
70°S	0.32	0.31	0.31	0.29	0.34	0.39	0.43	0.45	0.43	0.40	0.38	0.34
80°S	0.31	0.31	0.31	0.28	0.35	0.40	0.44	0.46	0.42	0.38	0.36	0.32
90°S	0.31	0.30	0.30	0.27	0.34	0.38	0.43	0.45	0.41	0.37	0.34	0.31

tained by Goetz *et al.* reduced to the new IGY scale [7] will be sufficient for our present purposes (Table 4.2).

The distribution of the water content is known for some radiosonde stations and may easily be extended to the whole aerological network. Unfortunately, water-content soundings are not in general as reliable as temperature and pressure soundings. At many points Hann's formula [8]

$$w = 0.25 \, e_{mb} \qquad\qquad (4.2)$$

must still be used. This formula is based on the extrapolation of the ground-level water vapour pressure e_{mb} and gives the total water content of the

TABLE 4.3. Total water content of the atmosphere, w (in cm precipitable water) as a function of latitude, season and altitude [9–11].

Mean values and mean limits are given; higher and lower values may be observed in exceptional circumstances.

Altitude (air pressure)		Latitude				
		0°	30°	45°	60°	70°
Warm or wet season						
1000 mb	mean	5.0	4.0	2.5	2.0	1.8
	min.	2.0	2.0	1.0	0.7	0.7
	max.	10.0	7.0	4.0	4.0	4.0
900 mb	mean	3.0	1.9	1.6	1.25	1.1
	min.	1.0	0.7	0.7	0.4	0.4
	max.	7.0	4.0	4.0	2.0	2.0
800 mb	mean	2.0	1.5	1.0	0.8	0.7
	min.	1.0	0.7	0.4	0.4	0.2
	max.	4.0	4.0	2.0	2.0	2.0
700 mb	mean	1.0	0.8	0.5	0.4	0.35
	min.	0.4	0.4	0.2	0.2	0.1
	max.	2.0	2.0	1.0	1.0	1.0
Cold or dry season						
1000 mb	mean	3.0	1.5	0.8	0.5	0.3
	min.	1.0	0.4	0.4	0.2	0.1
	max.	7.0	4.0	2.0	1.0	1.0
900 mb	mean	2.0	1.0	0.5	0.35	0.2
	min.	0.7	0.4	0.2	0.2	0.1
	max.	4.0	2.0	1.0	1.0	0.7
800 mb	mean	1.0	0.6	0.3	0.2	0.1
	min.	0.4	0.2	0.1	0.1	0.05
	max.	2.0	2.0	1.0	0.7	0.2
700 mb	mean	0.6	0.3	0.15	0.1	0.05
	min.	0.2	0.1	0.1	0.05	0.02
	max.	1.0	1.0	0.4	0.2	0.1

atmosphere, w, in centimetres of precipitable water. The formula may be subject to systematic errors of more than thirty percent at some places.

Table 4.3 shows the geographical distribution of atmospheric water vapour. It is based on the data published by Flohn [9], Valko [10] and Shands [11].

The turbidity coefficient B is known for only a very small number of locations. Ångström's formulae [1] for the geographical distribution of B are

$$B_\varphi = 0.040 + 0.085 \cos^2 \varphi \tag{4.3}$$

$$B_h = B_0 \exp(-0.0005\,h) \tag{4.4}$$

where φ is the geographical latitude and h is the altitude in metres.

These formulae are only very rough approximations and entirely neglect the considerable seasonal variations in B (particularly continental variations). Standard values of the turbidity coefficient for use in estimating the geographical distribution of global radiation are given in Table 4.4.

The intensity $I_{w,B,\gamma}$ is given in Diagram 4.2 (p. 150) as a function of γ and should be sufficiently accurate for interpolation to any value of w and B.

So far we have been concerned with direct solar radiation. Let us now consider the diffuse sky radiation which arrives at the earth's surface as a result of simple or multiple scattering (*i.e.* the so-called D-radiation). The scattering functions involved are extremely complicated and have only

TABLE 4.4. Standard values of the turbidity coefficient B as a function of latitude and altitude.

Altitude (air pressure)		70°N	60°N	45°N	30°N	0°	30°S	45°S	60°S	70°S
		Latitude								
1000 mb	Ångström	0.050	0.061	0.082	0.104	0.125	0.104	0.082	0.061	0.050
	lower limit	0.010	0.010	0.050	0.050	0.050	0.050	0.050	0.010	0.010
	upper limit	0.100	0.100	0.200	0.400	0.400	0.400	0.200	0.100	0.100
900 mb	Ångström	0.030	0.037	0.050	0.063	0.076	0.063	0.050	0.037	0.030
	lower limit	0.010	0.010	0.010	0.010	0.010	0.010	0.010	0.010	0.010
	upper limit	0.050	0.100	0.100	0.200	0.400	0.200	0.100	0.100	0.050
800 mb	Ångström	0.018	0.022	0.030	0.037	0.046	0.037	0.030	0.022	0.018
	lower limit	0.000	0.000	0.010	0.010	0.010	0.010	0.010	0.000	0.000
	upper limit	0.050	0.050	0.100	0.200	0.200	0.200	0.100	0.050	0.050
700 mb	Ångström	0.011	0.014	0.018	0.023	0.028	0.023	0.018	0.014	0.011
	lower limit	0.000	0.000	0.000	0.010	0.010	0.010	0.000	0.000	0.000
	upper limit	0.020	0.050	0.050	0.100	0.200	0.100	0.050	0.050	0.020

recently been evaluated numerically from Mie's theory [12] using special assumptions about the composition of the aerosols. The work of Deirmendjian [13] and Sekera and Ashburn [14] has shown that it is possible to estimate the total diffuse sky radiation D_γ on a horizontal plane with the aid of the following simple formula proposed earlier by Albrecht [15]:

$$D_\gamma = k_\gamma (I_{w,\gamma} - I_\gamma) \sin \gamma \qquad (4.5)$$

where k_γ is an empirical coefficient which in the Congo is given by

$$k_\gamma = 0.5 \sin^{\frac{1}{3}} \gamma \qquad (4.6)$$

$I_{w,\gamma}$ is the direct solar radiation in the absence of scattering by molecules and dust particles, and I_γ is the direct solar radiation when scattering effects are included.

The coefficient k_γ can be deduced either from multiple scattering theory with the albedo A of the ground taken into account (particularly for a low sun), or by direct observations for very different values of the turbidity and albedo. Table 4.5 gives the numerical data for the evaluation of $I_{w,\gamma}$ from the formula

$$I_{w,\gamma} = I_{m_{h1}w} - \Delta_{m_{h2}o_3} \qquad (4.7)$$

TABLE 4.5. Data for the evaluation of $I_{w,\gamma} = I_{m_{h1}w} - \Delta_{m_{h2}o_3}$. $m_{h1} \sim \sec z_\odot$; m_{h2} for 22 km is given in Table 3.1 (p. 51).

$m_{h1}w$ (cm)	$I_{m_{h1}w}$ (mcal · cm^{-2} · min^{-1})	$m_{h2}o_3$ (cm NTP)	$\Delta_{m_{h2}o_3}$ (mcal · cm^{-2} · min^{-1})
0.01	1948	0.20	14
0.02	1930	0.30	20
0.04	1910	0.40	26
0.07	1892	0.50	32
0.10	1880	0.60	37
0.20	1849	0.70	43
0.40	1810	0.80	48
0.70	1775	0.90	52
1.00	1753	1.00	57
2.0	1699	1.20	67
4.0	1641	1.40	76
7.0	1586	1.60	85
10	1546	1.80	93
20	(1474)	2.00	101
40	(1390)	2.50	120
70	(1310)	3.00	(140)
100	(1270)	4.00	(178)
		5.00	(213)

For a pure atmosphere without secondary scattering (and therefore independent of the albedo) the magnitude of k_γ should be 0.5. Scattering by dust leads to higher values of k_γ because forward scattering by dust is much stronger than backward scattering. However, the effect is considerably reduced by multiple scattering and it is possible that scattering by dust particles also involves some absorption. Therefore, in spite of the fairly high albedo in the Congo ($A=0.25$) the magnitude of k_γ for cloudless days and normal extinction ($\alpha=1.5$) is very nearly 0.5 for a high sun. On some days (sand dust, fine cirrus cloud, $\alpha=0.5$) much higher values of k_γ (0.67) are observed for a high sun. Such days appear to be exceptional, apart from stormy desert regions. On the other hand, it is clear that no formula for D_γ which is based on local conditions will be valid when $\gamma < 5°$ because the afterglow effects become more and more important towards sunrise and sunset.

Theoretically, the magnitude of the albedo exerts a considerable effect on k_γ, and therefore in any region where the albedo is different from $A=0.25$ the magnitude of k_γ should be adjusted accordingly. The reflected radiation is a fraction of the total G-radiation and will therefore be considered after we have considered the global radiation in the following paragraph.

The global radiation G is the sum of the vertical components of the direct solar radiation I and the diffuse sky radiation D. It is therefore given by

$$G_\gamma = I_\gamma \sin \gamma + D_\gamma \qquad (4.8)$$

Combining this with the formulae given above we have

$$G_\gamma = [(1-k_\gamma)I_\gamma + k_\gamma I_{w,\gamma}] \sin \gamma \qquad (4.9)$$

As noted above, this formula will only be valid when the albedo is $A=0.25$. In general, the contribution of the ground-reflected radiation to the multiply scattered diffuse sky radiation must be taken into account.

Ångström [16] demonstrated the influence of the albedo of the ground on the global radiation, particularly for a cloudy sky. The contribution of the ground-reflected radiation to the global radiation is

$$\int_0^\infty \int_0^{2\pi} \int_0^{\pi/2} A_{\varepsilon,A,\lambda} G_{A,\lambda} \sin \varepsilon \, k_\varepsilon \frac{I_{w,\varepsilon,\lambda}-I_{\varepsilon,\lambda}}{I_{w,\varepsilon,\lambda}} \, d\varepsilon \, dA \, d\lambda \qquad (4.10)$$

where $G_{A,\lambda}$ is the global incoming radiation with the additional effect of the albedo included, $A_{\varepsilon,A,\lambda}$ is the reflecting power (albedo) of the ground in the direction ε, A; k_ε is an empirical constant representing back scattering, $I_{w,\varepsilon,\lambda}$ is the direct radiation reduced only by absorption along the light path in the direction ε, and $I_{\varepsilon,\lambda}$ is the direct radiation at ground level when the sun is at an elevation ε.

It is clear that, in general, we know very little about the functions $A_{\varepsilon,A,\lambda}$, k_ε and $G_{A,\lambda}$, and it is therefore necessary to take mean values. Even when $I_{w,\varepsilon,\lambda}$ and $I_{\varepsilon,\lambda}$ are known for a large number of spectral regions, it is still of no great help to take account of the complete function because the other quantities in the equation are not adequately known. We shall therefore use the simplified expression

$$G_A = G + A \cdot G_A \sum_{\varepsilon=0}^{\pi/2} f_\varepsilon k_\varepsilon \frac{I_{w,\varepsilon} - I_\varepsilon}{I_{w,\varepsilon}} \qquad (4.11)$$

where f_ε represents the contribution of the sky region between ε and $\varepsilon + d\varepsilon$ to a horizontal receiver, $k_\varepsilon = k_\gamma$ for $\gamma = \varepsilon$, $I_{w,\varepsilon} = I_{w,\gamma}$ for $\gamma = \varepsilon$, and $I_\varepsilon = I_\gamma$ for $\gamma = \varepsilon$.

The ratio G_A/G for a black ground will be taken to be

$$G_A/G = \left(1 - A \sum_{\varepsilon=0}^{\pi/2} f_\varepsilon k_\varepsilon \frac{I_{w,\varepsilon} - I_\varepsilon}{I_{w,\varepsilon}} \right)^{-1} \qquad (4.12)$$

This formula will in general give an overestimate, especially when the albedo A has selective properties and is high at wavelengths for which the diffusivity is low (e.g. green vegetation, with low A at short wavelengths, and very high A in the near infrared). Nevertheless, Table 4.6 may give a good estimate of the influence of the ground albedo in special cases. Our empirical estimate of k_γ above a ground with $A = 0.25$ proves to give fairly good values for G_A on the continents, so long as there is no snow cover. On clear days the error will be small even over the surface of an ocean, but with a cloud cover the total radiation over the sea (under the same meteorological conditions) will be overestimated by about 7%.

Let us now consider the ratio of the diffuse sky radiation D_γ at solar altitude γ above a black ground and the diffuse radiation above a ground with an albedo A. Since in general G_A/G is independent of γ we can write

$$G_{\gamma,A} = (G_A/G)G_\gamma \qquad (4.13)$$

Combining this with Eq. (4.8), we have

$$D_{\gamma,A} = G_{\gamma,A} - I_\gamma \sin \gamma = (G_A/G)G_\gamma - I_\gamma \sin \gamma \qquad (4.14)$$

and

$$D_{\gamma,A}/D_\gamma = \frac{(G_A/G)G_\gamma - I_\gamma \sin \gamma}{G_\gamma - I_\gamma \sin \gamma} \tag{4.15}$$

In practice this ratio turns out to be near to the ratio indicated for a cloud in Table 4.6 and shows no dependence on B and γ, in spite of the very strong dependence of D_γ both on B and γ. The marked dependence of $D_{\gamma,A}/D_\gamma$ on γ reported by Sekera and Deirmendjian [17] for pure Rayleigh scattering is supported by Eq. (4.14). Careful ultraviolet observations of Bener [17a] for $0.15 < A < 0.85$ at Davos with $B \sim 0.05$ are in good agreement with Table 4.6 ($B = 0.4$), in accordance with the much higher scattering in the ultraviolet.

TABLE 4.6. Values of the ratio $G_A/G_{A=0.25}$ (compared with mean conditions, $A = 0.25$).

Conditions	Albedo (A)	Turbidity coefficient (B)					
		0.01	0.10	0.20	0.40	0.80	∞(cloud)
Black body	0.0	0.98	0.965	0.95	0.935	0.915	0.895
Sea	0.1	0.985	0.98	0.97	0.96	0.945	0.93
Grass	0.2	0.995	0.99	0.985	0.98	0.975	0.975
Desert	0.3	1.005	1.01	1.015	1.02	1.02	1.025
Normal snow	0.6	1.035	1.05	1.075	1.11	1.16	1.20
Fresh snow	0.8	1.05	1.08	1.125	1.18	1.27	1.37
Antarctic snow	1.0	1.07	1.12	1.18	1.27	1.38	1.56

A graphical method to evaluate D_γ or $D_{A,\gamma}$ is given in the appendix (Diagram 4.3, p. 154). The combined use of Diagrams 4.1, 4.2 and 4.3, together with Eq. (4.8) will allow estimates of I_γ, D_γ and G_γ for practical purposes. For low latitudes Tables 4.17–4.19 (p. 157) give daily sums of global radiation.

3. TOTAL RADIATION FALLING ON A TILTED PLANE

The geometrical problem has already been discussed in Chapter 2 and is resolved with the aid of the formula

$$\cos \beta = \sin \varepsilon \cos \gamma \cos (A_\odot - A_p) + \cos \varepsilon \sin \gamma \tag{4.16}$$

The influence of the tilt angle ε of the plane, and the angle β at which the direct radiation falls on the plane, on the diffuse sky radiation reaching the plane is extremely complicated and is very dependent on the cloudiness, the atmospheric turbidity and the albedo of the ground. Unfortunately, no published information appears to be available on this point. There exist many measurements of illuminance for a cloudless sky and different positions of the sun. These measurements indicate a much higher intensity near the sun as compared with other directions, and in general a much higher intensity near the horizon than in the zenith [18–21]. A theoretical calculation based on Mie's scattering function for a pure atmosphere gives much smaller differences [22], showing that in general a large number of dust particles exists in the atmosphere. A much better agreement between observation and theory has been obtained by Volz [23], Deirmendjian [13] and others who used Mie's scattering functions with theoretical dust compositions. Volz [23] points out that small differences in aerosol composition have a large effect on the sky luminance near the sun. However, the picture is still incomplete and integration over the region of the sky 'seen' by a tilted plane is still not possible.

A recent paper due to Kondratyev and Manolova [24] may give us a rough idea of the extent to which the anisotropy in the sky influences our estimates of the total radiation. These workers have reported measurements of diffuse and reflected radiation falling on pyranometers screened from the sun. Supposing that the albedo of the ground is $A = 0.15$ and the global radiation is $G = 1.0$ cal $\cdot$ cm^{-2} $\cdot$ min, we can eliminate reflected radiation and obtain the values given in Table 4.7.

TABLE 4.7. Diffuse sky radiation reaching a tilted plane compared with radiation reaching a horizontal plane: D_ε/D (sun at 40° altitude).

Angle between solar and plane azimuth	Tilt angle of the plane				
	0°	30°	45°	60°	90°
0°	1.00	1.25	1.28	1.30	0.86
90° and 270°	1.00	0.89	0.82	0.71	0.50
180°	1.00	0.67	0.56	0.53	0.36
$\cos^2 \varepsilon/2$	1.00	0.93	0.85	0.75	0.50
overcast sky $= \cos^3 \varepsilon/2$	1.00	0.90	0.78	0.65	0.36

It is clear that the sky near the sun has a higher luminance than at 90°, or even at 180°, from the sun. This becomes even more apparent if we consider the angular distance between the sun and the normal to the tilted plane. Kondratyev and Manolova reported that because of the relatively low intensity of the diffuse radiation, the approximation based on the $\cos^2 \varepsilon/2$ law may in general be sufficient for global radiation. Thus,

$$D_\varepsilon = D_\gamma \cos^2 \frac{\varepsilon}{2} = k_\gamma (I_{w,\gamma} - I_\gamma) \sin \gamma \cos^2 \frac{\varepsilon}{2} \qquad (4.17)$$

Moreover, Kondratyev and Manolova reported that for an overcast sky there is no azimuthal anisotropy. However, Moon and Spencer [25] and many others have reported a pronounced anisotropy with the elevation angle as indicated in the last line of Table 4.7. The $\cos^2 \varepsilon/2$ law is obtained if one integrates over the entire hemisphere above the plane, with ground reflection ignored.

Tonne and Normann [26] have given tables for the mean distribution of luminance over the sky as a function of cloudiness. Their results are based on all available information and are given in Table 4.8. The results of an integration of the data given in Table 4.8 for a vertical plane are shown in Table 4.9. The values given in this table are in reasonable agreement with

TABLE 4.8. Distribution of luminance over the sky as a function of cloudiness (sun at 20° altitude) [26].

Cloudiness	Vertical through the sun						Opposite vertical					
	0°	15°	30°	45°	60°	75°	75°	60°	45°	30°	15°	0°
0	15	18	11	5	2.5	1.5	0.75	0.75	0.85	1.2	2.0	2.7
cloudy	4	7	5	2.5	1.7	1.3	0.90	0.88	0.95	1.1	1.3	1.4
50%	3	5	3.5	2.0	1.5	1.2	0.92	0.90	0.92	0.95	0.98	1.05
75%	1.5	1.9	1.8	1.3	1.2	1.05	0.95	0.92	0.90	0.82	0.75	0.65
100%	0.4	0.55	0.7	0.85	0.95	0.99	0.99	0.95	0.85	0.70	0.55	0.40

TABLE 4.9. Percentage of diffuse radiation D_γ falling on vertical planes as a function of the elevation γ of the sun and the azimuthal angle difference $A_\odot - A_p$ for mean cloudiness [26].

Solar elevation	$A_\odot - A_p$						
	0°	30°	45°	60°	90°	135°	180°
10°	85	80	75	62	55	50	55
20°	82	78	73	62	50	45	50
30°	80	76	71	61	50	40	40
45°	75	71	66	57	49	38	38
60°	55	53	48	46	43	38	38
75°	43	42	42	41	40	38	38
90°	36	36	36	36	36	36	36

Kondratyev's observations for $\varepsilon = 90°$ and $A_\odot - A_p = 90°$, $180°$ and $270°$. For $A_\odot - A_p = 0°$ there is also good agreement at solar altitudes between $10°$ and $20°$. The data in Tables 4.8 and 4.9 are, of course, valid only for mean-turbidity conditions and those given in Table 4.9 are further limited to mean cloudiness and vertical planes. However, these results provide a reasonable representation of the problem.

There appears to exist a possibility of taking these anisotropies into account, at least qualitatively, by reducing the luminance of the clear fraction of the sky by a factor of 0.75 and adding the difference to the direct component in such a way that the same global radiation is obtained on a horizontal plane. A factor $\cos \varepsilon/2$ is then added to represent the cloudy fraction of the sky, and in this way the difference between clear and cloudy sky is also allowed for. The difference between the total radiation on a tilted plane computed in this way and the value found for an isotropic sky will help us to discuss the accuracy of our theoretical estimates.

A further complication is the fact that a tilted plane also faces a part of the ground and it is necessary to calculate the radiation reflected from the ground. This component is essentially proportional to the global radiation G_γ and the albedo A' of the ground facing the plane. A' may be considerably different from the mean albedo A. Assuming diffuse reflection, the reflected radiation is given by

$$R_\varepsilon = G_\gamma A' \left(1 - \cos^2 \frac{\varepsilon}{2}\right) \tag{4.18}$$

Specular reflection (water, mirrors, etc.) may give much higher values on a plane facing a low sun. The specular component is given by

$$S_{\varepsilon,\text{mirror}} = I_\gamma(\sin \beta + A' \sin \beta') \tag{4.19}$$

where

$$\sin \beta' = \cos \varepsilon \sin \gamma - \sin \gamma \cos \gamma \cos (A_\odot - A_p) \tag{4.20}$$

The following equations again consider only the case of a diffuse reflecting ground:

$$G_{\gamma,\varepsilon,A,A'} = I_{\gamma,\varepsilon} + D_{\gamma,\varepsilon,A} + R_{\varepsilon,A,A'} =$$
$$= I_\gamma \sin \beta + D_{\gamma,A} \cos^2 \frac{\varepsilon}{2} + G_{\gamma,A,A'} \left(1 - \cos^2 \frac{\varepsilon}{2}\right) \tag{4.21}$$

(for clear sky conditions).

If the anisotropy deduced from Kondratyev's observations [24] is taken into account then the total radiation is given by

$$
G_{\gamma,\varepsilon,A,A'} = \left(I_\gamma + \frac{0.25}{\sin \gamma} D_{\gamma,A,\text{blue}} \sin \beta \right) + \left(0.75 \, D_{\gamma,A,\text{blue}} \right.
$$
$$
\left. + \cos \frac{\varepsilon}{2} D_{\gamma,A,\text{cloud}} \right) \cos^2 \frac{\varepsilon}{2} + G_{\gamma,A,A'} \left(1 - \cos^2 \frac{\varepsilon}{2} \right) \tag{4.22}
$$

where $D_{\gamma,A,\text{blue}}$ is the diffuse radiation coming from a cloudless region of the sky and $D_{\gamma,A,\text{cloud}}$ is the diffuse radiation coming from the clouds. These expressions will be discussed again below.

4. THE INFLUENCE OF ATMOSPHERIC OZONE ON GLOBAL RADIATION

It has already been noted that as far as global radiation is concerned, absorption by ozone may be regarded as fairly small. Nevertheless, the latitudinal and seasonal variation given in Table 4.2 will be used to compute a correction to the values for a 0.34 cm NTP ozone layer. The correction to be applied to the direct radiation is tabulated in Tables 4.10a and 4.10b.

TABLE 4.10a. Correction (in mcal · cm^{-2} · min^{-1}) per 0.10 cm ozone, to be applied to direct radiation I_γ in the case of an ozone content $\neq$ 0.34 cm NTP (IGY scale).

m_γ	B							
	0.000	0.025	0.050	0.075	0.100	0.200	0.400	0.800
1	−7	−7	−7	−6	−6	−4	−3	−2
2	−8	−8	−7	−6	−6	−4	−2	−1
4	−14	−13	−11	−9	−8	−4	−1	0
6	−18	−16	−14	−11	−8	−4	−1	0
10	−23	−19	−15	−11	−7	−2	0	0

TABLE 4.10b. Correction (in %) per 0.10 cm ozone, to be applied to direct radiation I_γ in the case of an ozone content $\neq$ 0.34 cm NTP (IGY scale).

m_γ	B				
	0.000	0.100	0.200	0.400	0.800
1	−0.65	−0.60	−0.55	−0.45	−0.35
2	−0.85	−0.80	−0.75	−0.60	−0.45
4	−1.55	−1.45	−1.15	−0.85	(−0.65)
6	−2.25	−2.05	−1.65	−1.20	(−0.95)
10	−3.40	−2.80	−2.55	(−2.20)	(−1.80)

5. THE INFLUENCE OF CLOUDS ON GLOBAL RADIATION

The only region where the influence of clouds on solar radiation may be ignored is the desert. In any estimate of available solar energy the study of clear-weather conditions will only give the maximum amount of energy reaching a horizontal receiver. Unfortunately, the mean solar radiation for cloudy weather has not attracted the interest which it merits, so that there is a considerable lack of detailed information.

Clouds are complex phenomena, varying from thin, transparent cirrus, exerting little influence on global radiation, to thick and dark thunderstorm clouds which may reduce the global radiation to 1% of its normal value for some hours. The brightness of the same cumulus cloud may greatly vary, depending on its position with regard to the sun. There are but few stations at which each cloud layer is separately observed, and at which both the type and extent of every layer is noted. In all other locations the estimated values of total cloudiness are of very little help in the present problem.

As was already pointed out in a study of spherical radiation [27], the only possible solution is to take the duration of sunshine, n, as a measure of cloudiness, particularly as this type of information does take into account the optical density of the clouds at a certain stage. Unfortunately, the optical thickness of the same cloud appears much greater with a low than with a high sun; thus Robitzsch [28] indicates a dependence on cot γ.

However, in spite of these difficulties, let us consider a few general principles which may help us to establish, in a general way, the influence of clouds on global radiation. To do this let us write down the expression for the global radiation in the following form

$$G_{\gamma, n/N} = I_\gamma \sin \gamma \, (1 - C) + D_{\gamma, \text{blue}} + D_{\gamma, \text{cloud}} \qquad (4.23)$$

in which

$$C = 1 - n/N$$

$$D_{\gamma, \text{blue}} = D_\gamma (1 - C) = D_\gamma \cdot n/N$$

$$F_{\gamma, \text{cloud}} = G_\gamma \cdot f(n/N)$$

where $D_{\gamma, \text{blue}}$ is the diffuse radiation of the cloudless region of the sky, $D_{\gamma, \text{cloud}}$ is the diffuse radiation from the cloudy regions, and C is the cloudiness, following from the ratio between the observed duration of sunshine n and the maximum possible value N for the same date. The best estimate of N

is the highest observed value at a particular place and date over many years. If a long series of observations is not available, then the time during which the sun is higher than 3° above the horizon can be chosen, provided there are no obstructions above the optical horizon [29].

The first term in Eq. (4.23) requires no explanation, although Sauberer [30] has in fact tested this assumption for Vienna, and has observed lower intensities of direct radiation during periods of intermittent sunshine for which most observers overestimate n [31]. If the influence of the cloud albedo and shade on the luminance can be neglected, then the relation for $D_{\gamma,\text{blue}}$ should be the same as for the direct component. There remains the third term in the equation for which no simple assumptions are valid. Sauberer and Dirmhirn [32] and Ångström [33] observed the highest value of $D_{\gamma,\text{cloud}}$ at $C=0.7$ and much lower values again for $C=1.0$. Schüepp [34] has reported the same value for $D_{\gamma,\text{cloud}}$ for the equatorial region at $C=0.7$ and $C=1.0$. Shieldrup-Paulsen [35] appears to have been the first to give a physical explanation of this result, attributing it to the screening of the brilliant cloud tops in the background by the dark base of clouds in the foreground.

Records of sky radiation at Leopoldville, collected over a 10-year period, show that generally direct radiation and the fraction of blue sky radiation $D_{\gamma,\text{blue}}$ are reduced by a factor $< n/N$ because cirrus and other transparent clouds do not affect n. In this case $D_{\gamma,\text{cloud}}$ is increased: $D_{\gamma,\text{cloud}} > G_{\gamma} \cdot f(n/N)$, but this difference concerns particularly the sector of the sky near the sun. To estimate the radiation falling on inclined planes, it is better to use the formula as it is developed here, which gives low values for $D_{\gamma,\text{cloud}}$ on a horizontal plane, but is correct for the sky opposite to the sun. The global radiation for the same period agrees within 0 to 5 % with the estimated values.

Provided the analysis is limited to global radiation falling on a horizontal plane, we can use an empirical formula of the Ångström type [36]:

$$G_C = G_0 [\alpha + (1-\alpha) \cdot n/N] \tag{4.24}$$

or the second order formula:

$$G_C/G_0 = a + [b(n/N)^2 + c(n/N)]^{\frac{1}{2}} \tag{4.25}$$

In the second formula the ratio c/b describes the degree of curvature. The special case $c=0$ leads us back to the Ångström formula with $a=\alpha$ and $b=1-\alpha$. On the other hand, $c/b=1.8$ is a typical value where the blue sunshine cards of the Campbell–Stocks type recorder (see p. 276) are evaluated by carefully following the instructions of the revised WMO guide [37], which include reasonable reductions for narrowing in the trace.

In the case of tilted surfaces it is necessary to use the general equation Eq. (4.23) and a more exact value must be taken for $D_{\gamma,cloud}$.

TABLE 4.11. The function $f(n/N)$ for a number of stations deduced from Eqs. (4.23) and (4.25) for daily sums [38–40].

n/N	Place*			
	Arosa	Base Baudoin	Congo	Potsdam
1.0	0.00	0.00	0.00	0.00
0.9	0.06	0.09	0.05	0.04
0.8	0.11	0.18	0.10	0.08
0.7	0.17	0.26	0.15	0.12
0.6	0.22	0.32	0.20	0.16
0.5	0.27	0.38	0.25	0.19
0.4	0.32	0.44	0.29	0.22
0.3	0.36	0.50	0.34	0.25
0.2	0.40	0.55	0.37	0.26
0.1	0.42	(0 59)	0 39	0.26
0.05	0.42	(0.61)	0.39	0.24
0.0	0.35	(0.64)	0.32	0.14

* Arosa, Switzerland: 47°N, 1800 m; Base Baudoin, Antarctica: 70°S; Congo: stations between 0° and 12°S and 0 and 1650 m; Potsdam, Germany: 52°N, 60 m.

Table 4.11 gives the evaluation of $f(n/N)$ for several stations. These data were evaluated for daily sums. No observations for the hourly sums appear to be available. The Potsdam data compare well with the observations of Sauberer and Dirmhirn [32] in Vienna, and may be looked upon as representative for low-level stations at middle latitudes. On the other hand, the Congo data are in satisfactory agreement with the results from high altitude stations at middle latitude (Arosa). It is not as yet possible to link these two types of climate by a quantitative formula, and no data appear to have been published for stations where conditions intermediate between these two types are observed. At Arctic stations the function $f(n/N)$ seems to be higher owing to the high albedo of the snow cover [40]. This is a considerable difficulty as far as the general problem is concerned, because the solution of the problem should be independent of geographical considerations. Moreover, there are many stations at which the methods of computation used with sunshine recorders are quite different. This has been shown by the

Working Group on Solar Radiation in Region I [41]. Other stations use different types of sunshine recorders which have not been compared with the Campbell–Stocks recorders (for example, stations in the USA, the French Territories before the IGY, and so on).

It is not essential to have a concise formula for the relation mentioned above; it will be sufficient to have some representative values. Nevertheless, it is clear that different results will be obtained when this function is used with the monthly mean values of the cloudiness. In fact, this is the weakness of non-linear relations which are very awkward to use. However, when reasonable information is available on the frequency of each type of cloudiness, an alternative method may be used in which the weighted mean G_C of all the observations is formed with daily mean cloudiness equal to 0, 0.1, 0.2, ..., 1.0:

$$G_C = (n_0 \, G_0 + n_{0.1} \, G_{0.1} + \ldots + n_1 \, G_1) \Big/ \sum_{v=0}^{1} n_v \qquad (4.26)$$

The frequency distribution of n/N as a function of the mean monthly value of n'/N' is of course different at different points. Table 4.12, derived from the data for the Congo, can be used to obtain fairly good estimates of $f'(n'/N')$ for the mean monthly values. Similar tables have been prepared for the USA by Liu and Jordan [42]. This frequency distribution, when used with Eq. (4.25) and Table 4.11 gives the approximate values of $f'(n'/N')$

TABLE 4.12. Frequency distribution of daily n/N as a function of the monthly mean value n'/N'.

Monthly n'/N'	Daily n/N											
	0.00	0.05	0.10	0.20	0.30	0.40	0.50	0.60	0.70	0.80	0.90	1.00
1.00	0.00	0.00	0.00	0.00	0.00	0.00	0.00	0.00	0.00	0.00	0.00	1.00
0.90	0.00	0.00	0.00	0.01	0.02	0.02	0.02	0.04	0.05	0.05	0.14	0.65
0.80	0.01	0.02	0.01	0.02	0.02	0.03	0.03	0.05	0.10	0.15	0.30	0.25
0.70	0.02	0.02	0.03	0.05	0.05	0.05	0.05	0.08	0.15	0.22	0.15	0.15
0.60	0.03	0.02	0.04	0.07	0.05	0.08	0.12	0.13	0.15	0.12	0.10	0.08
0.50	0.06	0.04	0.05	0.10	0.10	0.13	0.15	0.10	0.10	0.08	0.05	0.04
0.40	0.08	0.06	0.06	0.10	0.15	0.15	0.15	0.07	0.06	0.06	0.04	0.02
0.30	0.12	0.12	0.08	0.14	0.15	0.13	0.10	0.06	0.05	0.03	0.02	0.00
0.20	0.20	0.15	0.10	0.15	0.20	0.10	0.05	0.03	0.02	0.00	0.00	0.00
0.10	0.40	0.20	0.15	0.10	0.05	0.05	0.05	0.00	0.00	0.00	0.00	0.00
0.05	0.70	0.10	0.10	0.07	0.03	0.00	0.00	0.00	0.00	0.00	0.00	0.00
0.00	1.00	0.00	0.00	0.00	0.00	0.00	0.00	0.00	0.00	0.00	0.00	0.00

TABLE 4.13. The function $f'(n'/N')$ for a number of stations for application to monthly mean values.

Monthly n'/N'	Place		
	Congo	Potsdam	Arctic snow*
1.00	0.00	0.00	0.00
0.90	0.05	0.04	0.09
0.80	0.10	0.08	0.18
0.70	0.15	0.12	0.24
0.60	0.20	0.15	0.31
0.50	0.25	0.17	0.37
0.40	0.28	0.19	0.43
0.30	0.31	0 21	0.48
0.20	0.33	0.22	0.52
0.10	0.34	0.20	0.53
0.05	0.34	0.18	0.50
0.00	0.32	0.14	0.47

* Data from Congo reduced to albedo $A = 1$.

shown in Table 4.12, when applied to mean monthly values. These results are quite different from those in Table 4.11 for n'/N' between 0.05 and 0.40. When only the mean monthly data are available, Table 4.13 must be used instead of Table 4.11 for the daily sums.

The effect of cloudiness on radiation falling on tilted planes is very dependent on the position of the sun. Such calculations are particularly useful when one is interested in finding a relatively stable source of energy rather than particularly high maximum values. The radiation from clouds is much more sensitive to the albedo A of the underlying ground than the radiation from the clear sky. This has already been shown in Table 4.6. It is not clear whether the values of the coefficient k_y obtained empirically for a clear sky may also be used with clouds. Perhaps much depends on the characteristics of the cloud cover, including its height above ground. For distant clouds some absorption effects may be as important as for the clear sky. On the other hand, in the case of mist, absorption effects may in general be entirely neglected. In the case of an overcast sky, the direct radiation will not penetrate to the earth's surface, and the general formula Eq. (4.12) will become

$$G_A/G = \left(1 - \sum_{\varepsilon=0}^{\pi/2} f_\varepsilon k_\varepsilon\right)^{-1} \tag{4.27}$$

As explained above, k_ε may be much closer to 0.5 than in the case of a cloudless atmosphere. Petherbridge [43] observed the sky to be much brighter near the horizon over snow than over dark ground. In the special case $k_\varepsilon = 0.5$ we have $G_A/G = 2$, which is the maximum possible effect of the albedo in a perfectly diffusing medium. When viewed from above, such a cloud should have an albedo of $A = 1$, while the mean value observed over a continent is of the order of 0.5. Fritz [44] has emphasized that the albedo of the cloud cannot be separated from the albedo of the underlying ground (or intermediate cloud layers).

In the case of a broken, thin cloud cover, the luminance of the cloud is much greater near the sun than in other directions. Cumulus clouds are darkest when just in front of the sun, and have a very bright borderline, although as a whole they appear brightest when opposite to the sun.

Very little is known about the statistical mean luminance of a cloud as a function of the angular distance from the sun. Eq. (4.22) does, to some extent, take this effect into account. Only careful measurements will bring us nearer to a solution of this problem. In any case, the problem of clouds can only be solved by statistical methods, and individual cases may depart considerably from average conditions.

6. RADIATION FALLING ON A TILTED PLANE IN CLOUDY WEATHER

It may be useful to write down the complete formula for radiation in cloudy weather by substituting Eqs. (4.23)–(4.27) into Eqs. (4.21)–(4.22). The result of this substitution for the case of an isotropic sky and daily sums is

$$G_{\gamma,A,A',\varepsilon,n/N} = I_{\gamma,\varepsilon,n/N} + D_{\gamma,\varepsilon,A,n/N} + R_{\gamma,A,A',n/N} \tag{4.28}$$

where

$$I_{\gamma,\varepsilon,n/N} = I_\gamma(n/N) \sin \beta \tag{4.29}$$

$$D_{\gamma,\varepsilon,A,n/N} = [D_{\gamma,A}(n/N) + G_{\gamma,A}f(n/N)] \cos^2 \frac{\varepsilon}{2} \tag{4.30}$$

$$R_{\varepsilon,A,A',n/N} = G_{\varepsilon,A,n/N} \cdot A' \cdot \left(1 - \cos^2 \frac{\varepsilon}{2}\right) \tag{4.31}$$

For practical application Eq. (4.28) may be written in terms obtainable entirely from available formulae and diagrams:

$$G_{\gamma,\varepsilon,A,A',n/N} = I_\gamma \left[\sin\beta \cdot (n/N) + \sin\gamma \cdot f(n/N) \cos^2 \frac{\varepsilon}{2} \right.$$

$$\left. + \{(n/N) + f(n/N)\} \cdot A' \cdot \sin\gamma \cdot \left(1 - \cos^2 \frac{\varepsilon}{2}\right) \right]$$

$$+ D_{\gamma,A}[(n/N) + f(n/N)] \cdot \left[\cos^2 \frac{\varepsilon}{2} + A' \cdot \left(1 - \cos^2 \frac{\varepsilon}{2}\right)\right] \quad \text{(4.28 bis)}$$

where

γ is the elevation of the sun as given by Diagram 4.1 (p. 148)

ε is the tilt angle of the receiver

A' is the albedo in front of the receiver

A is the general albedo of the region

$n/N = 1 - C$ is the relative duration of sunshine

β is the angle of incidence on the receiver as given by Eq. (4.16)

$I\gamma$ is the direct solar radiation, which (when not measured) may be derived from Diagram 4.2 (p. 150)

$D_{\gamma,A}$ is the scattered radiation from the sky, which (when not measured) may be derived from Diagrams 4.3 and 4.4 (p. 154 and p. 156, resp.)

$f(n/N)$ for daily sums is given in Table 4.11.

In the case of an anisotropic sky and daily sums the result of the above-mentioned substitution is

$$G_{\gamma,\varepsilon,A,A',n/n} = I_\gamma (n/N) \left[\sin\beta + A' \cdot \sin\gamma \cdot \left(1 - \cos^2 \frac{\varepsilon}{2}\right) \right]$$

$$+ D_{\gamma,A}(n/N) \left[\frac{0.25 \sin\beta}{\sin\gamma} + A' + (0.75 - A') \cos^2 \frac{\varepsilon}{2} \right]$$

$$+ G_{\gamma,A} f(n/N) \left[A' + \left(\cos \frac{\varepsilon}{2} - A'\right) \cos^2 \frac{\varepsilon}{2} \right] \quad \text{(4.32)}$$

For mean monthly values the data for $f(n/N)$ from Table 4.11 should be replaced by $f'(n'/N')$ from Table 4.13 in all three Equations (4.28), (4.28a) and (4.32). Nevertheless quite large errors may still be introduced.

7. EXAMPLES

We are now in a position to apply the foregoing considerations to extreme and average conditions at any point on the earth with allowance for the daily and seasonal variations and the degree of cloudiness. Although the problem is much too complicated to allow a general solution, nevertheless the various tables and formulae are sufficient for normal calculations in so far as the meteorological conditions at a particular location are known.

It is outside the scope of this book to give a general description of the way in which radiation falling on tilted planes depends on the latitude, season, time, turbidity, cloudiness and albedo. The number of variables involved is far too large. The examples which we shall consider are given below in graphical form and refer principally to an equatorial station and low-turbidity conditions at the equinox. One of the reasons for this choice is that under these conditions the calculations become easier than otherwise.

Fig. 4.1 shows the mean variation of direct (I), diffuse (D) and global (G) radiation with the elevation of the sun. These observational data were obtained at Stanleyville ($\varphi=0°$, $\Lambda=25°E$, $h=450$ m) in 1952–1953 and compare quite well with calculated values (Diagram 4.3, p. 154). The quantities B, w and p indicated in the caption are the turbidity coefficient, the water content and the pressure respectively. Such curves may easily be obtained for any location for which estimates of the turbidity and humidity are available.

Fig. 4.2 shows the mean daily variation in the radiation on clear days in June and December 1958 at Leopoldville ($\varphi=4°S$, $\Lambda=15°E$, $h=450$ m). June corresponds to a dry hazy period when the humidity is lower, and the turbidity is still greater, than at Stanleyville in the case of the broken curves of Fig. 4.1. December corresponds to a humid and clear period when the turbidity is even lower than in the case of the continuous curves for Stanleyville.

It is clear that the I-radiation is greatly reduced by haze, while the diffuse D-radiation becomes much stronger. Some of the scattered radiation is lost to space and hence the global G-radiation is also reduced in spite of the high contribution due to the diffuse sky radiation. Between June and December three other factors must also be taken into consideration: (1) the change in the sun–earth distance is such that the intensity outside the atmosphere is higher by 7% in December than in June; (2) the elevation of the sun at noon for $\Lambda=4°S$ is 60° in June and 71° in December, which gives a

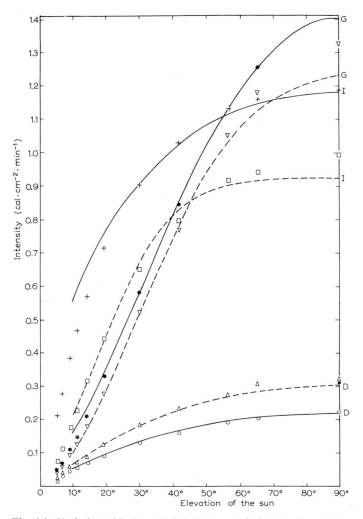

Fig. 4.1. Variation of I-, D- and G-radiation with the elevation of the sun at Stanleyville.
The curves show the observed values:
——— for the months II, III, IV, V, IX, X, XI, XII ($B = 0.100$, $w = 5.0$ cm, $p = 960$ mb).
- - - - - for the months I, VI, VII, VIII ($B = 0.250$, $w = 4.0$ cm, $p = 960$ mb).
The symbols show the calculated values:
● corresponds with G ———; ▽ corresponds with G - - - - -
+ corresponds with I ———; □ corresponds with I - - - - -
○ corresponds with D ———; △ corresponds with D - - - - -

References pp. 159–160

difference of 5 % in sin γ; (3) the effect due to the turbidity is still one-half of the total observed difference of 25 % in the G-radiation.

Fig. 4.3 shows the seasonal variation of the daily sum of G-, D- and

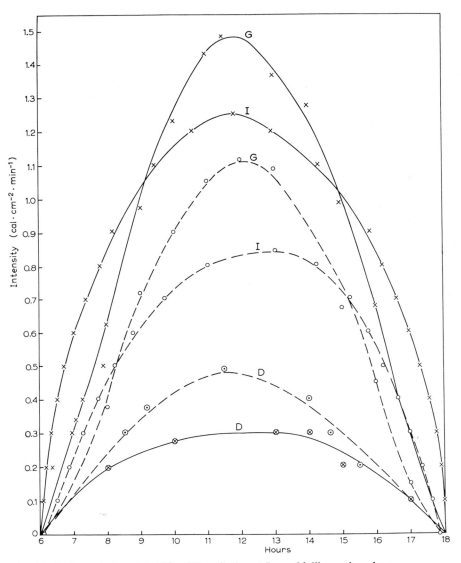

Fig. 4.2. Daily variation of I-, D- and G-radiation at Leopoldville on clear days.
———— and × : December 1958
- - - - - and ○ : June 1958

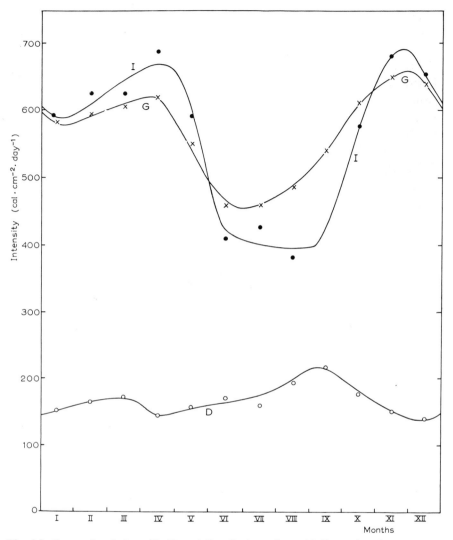

Fig. 4.3. Seasonal variation of I-, D- and G-radiation at Leopoldville on clear days.

I-radiation at Leopoldville on clear days. A combination of the three effects listed under Fig. 4.2 gives rise to a rather complicated form of these three curves.

The seasonal variation in the intensity of solar radiation on clear days can only be estimated from a knowledge of both meteorological and astronomical conditions. Moreover, conditions on clear days may prove to be of very little value in humid regions where absolutely clear days are exceptional.

Figs. 4.4, 4.5 and 4.6 take cloudiness into account. These isopleths of the seasonal variation of the mean monthly daily course of radiation are based on IGY observations at Leopoldville in 1958 and are quite different from what could be derived from the clear-weather results of Figs. 4.2 and 4.3. It is evident that considerable differences are observed from year to year . but there is a definite general trend.

The daily maxima between 13 h 30 m and 14 h 30 m in the I-radiation (Fig. 4.4) are a typical feature of the west coast of the continent, with the prevailing low stratus cloud in the morning. Maximum of low-level clouds is observed between 9 and 10 hours and minimum of low- and mean-level clouds generally occurs at 17 hours. High-level clouds with their afternoon

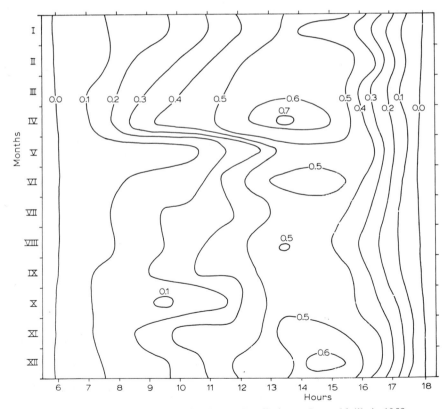

Fig. 4.4. Isopleths (cal · cm^{-2} · min^{-1}) of mean I-radiation at Leopoldville in 1958.

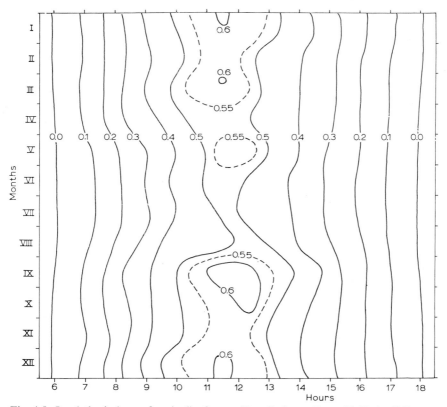

Fig. 4.5. Isopleths (cal · cm^{-2} · min^{-1}) of mean D-radiation at Leopoldville in 1958.

maximum are of little consequence. The seasonal variation in 1958 is not representative; only the maxima in April and November and the minimum in July appear fairly regularly every year.

The D-radiation isopleths (Fig. 4.5) are quite different. In fact, the diffuse radiation is in general higher in cloudy weather than on clear days; therefore a maximum of D-radiation is observed at 11 h 30 m. The maximum daily sums appear generally in March and September and the minimum in July.

The G-radiation isopleths (Fig. 4.6) exhibit great fluctuations in the seasonal variation from year to year. Maxima normally prevail in March–April and November, and minima in June and July. The seasonal variation was therefore less accentuated than is normally the case. It is evident that such details, and especially the amplitude of interannual variations, cannot

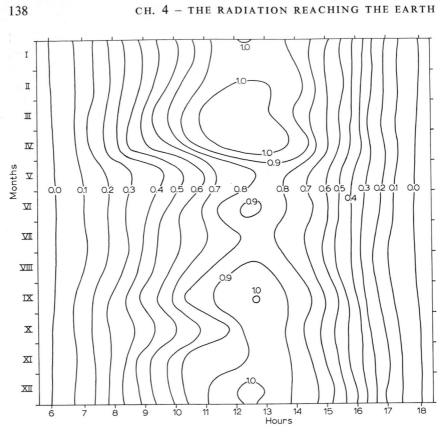

Fig. 4.6. Isopleths (cal $\cdot$ cm^{-2} $\cdot$ min^{-1}) of mean G-radiation at Leopoldville in 1958.

be determined theoretically. The maximal G-radiation is observed at 12 h 30 m.

There exist very few examples which show clearly the effect of the albedo A of the ground (over a large area) on global radiation. Dr. I.Dirmhirn has kindly allowed us to reproduce the results of observations carried out at the Patscherkofel station in Austria (47°N, 13°E, 2000 m above sea level) (Fig. 4.7). These results indicate the considerable effect of albedo, particularly for high cloudiness. The mean solar declination δ is not the same in April and in September, so that April data are higher by a few percent. On the other hand, snow cover may already be broken in April and the albedo may thus be lower than in winter. The two effects act in opposite directions and may tend to cancel each other out. Under clear-sky conditions the effect is found to be higher by 11% than the calculated result (Table 4.6). The mutual cancellation of the two effects mentioned above is thus probably insufficient. The higher the cloudiness, the larger the difference between the two months, in accordance with the last line of Table 4.6.

There are very few observations engaged in measuring the intensity of

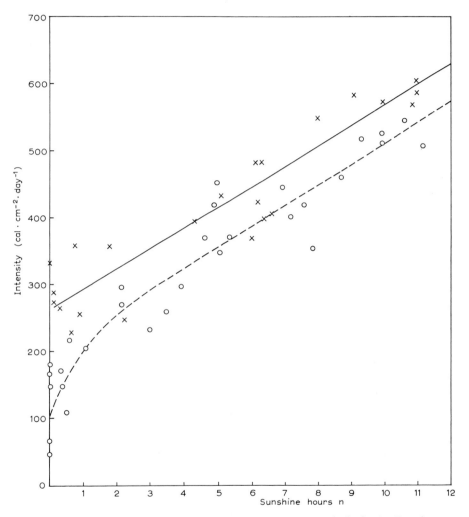

Fig. 4.7. Dependence of G-radiation on duration of sunshine n and albedo A at Patscherkofel.
——— and × : April 1960 (snow), $A = 0.6$.
- - - - - and ○ : September 1960 (grass), $A = 0.2$.

radiation on vertical or tilted planes. Generally speaking, there is no information about the albedo of the ground facing the recording device. Furthermore, no information is available about the obstructions of the horizon. The following Figures will illustrate the results of numerical calculations based on Eqs. (4.28)–(4.32) and (4.18); no comparison is made with experiment. It is hoped that these calculations will stimulate interest in such measurements.

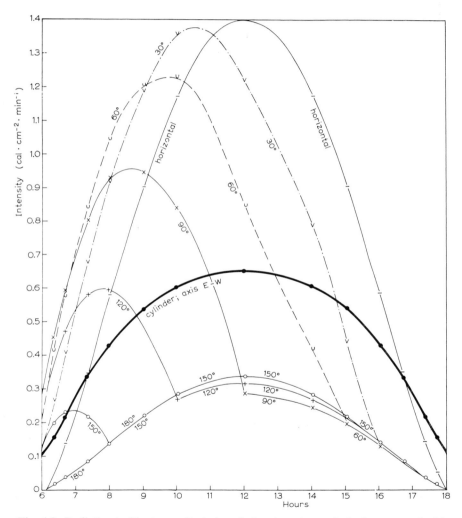

Fig. 4.8. Radiation incident on a tilted plane facing the solar vertical, above ground with diffuse albedo $A' = 0.25$ (equator, equinox, clear days).

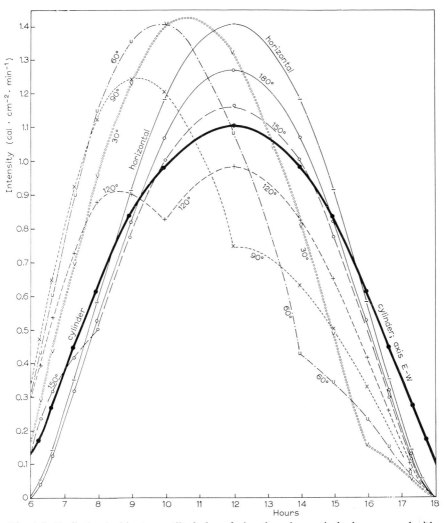

Fig. 4.9. Radiation incident on a tilted plane facing the solar vertical, above ground with diffuse albedo $A' = 0.90$ (equator, equinox, clear days).

Four cases will be distinguished, namely,

(a) Uniform sky (intensity independent of height above the horizon or azimuth); diffuse albedo of the ground in front of the receiver $A'=0.25$. Receiver does not shade the ground.

(b) Uniform sky; diffuse ground albedo; $A'=0.90$. Receiver does not shade the ground.

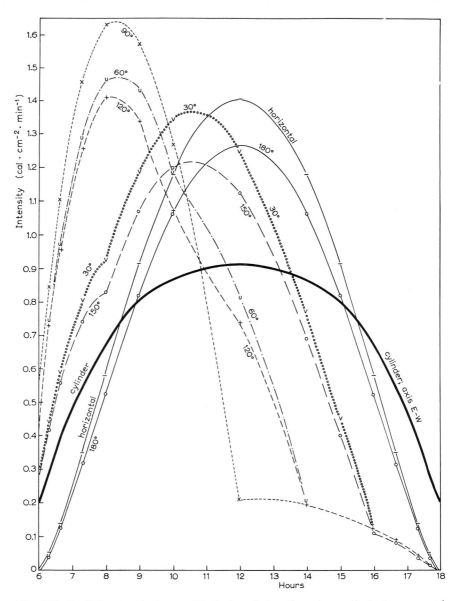

Fig. 4.10. Radiation incident on a tilted plane facing the solar vertical, above ground with specular albedo $A' = 0.90$ (equator, equinox, clear days).

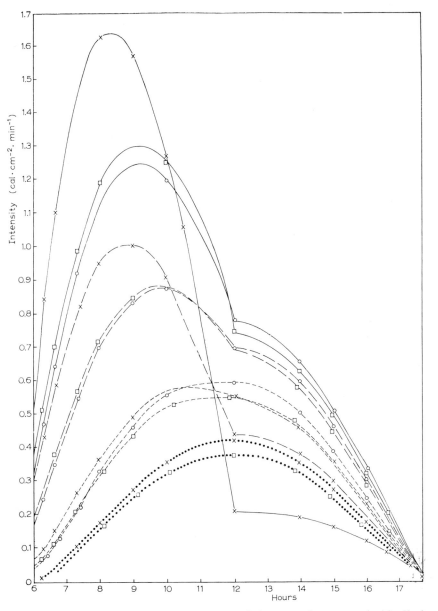

Fig. 4.11. Radiation incident on a vertical plane facing east, above ground with albedo $A' = 0.90$ (equator, equinox, clear days).

———— $n/N = 1$ ○ Diffuse reflection, $A' = 0.90$, isotropic sky
– – – $n/N = 0.5$ × Specular reflection, $A' = 0.90$, isotropic sky
- - - - $n/N = 0.1$ □ Diffuse reflection, $A' = 0.90$, anisotropic sky
...... $n/N = 0$

(c) Uniform sky; specular ground albedo; $A'=0.90$. Receiver does not shade the ground.

(d) Anisotropic sky (Eq. (4.32)); diffuse ground albedo; $A'=0.90$. Receiver does not shade the ground.

Fig. 4.8 shows the intensity of solar radiation on an inclined plane for case (a) as defined above. Cases (b) and (c) are illustrated by Figs. 4.9 and 4.10 respectively. Results are also shown for the global radiation falling on a cylinder with its axis normal to the sun's 'orbit' (example of low-cost solar water heater).

Fig. 4.11 shows a comparison between cases (b), (c) and (d). The effect of the nature of the ground is very considerable, but somewhat similar curves are obtained when the anisotropy of the sky above a diffuse reflecting ground is considered. With a very hazy atmosphere the effect of anisotropy may be three times larger as compared with the results of Fig. 4.11. However, even then it is of secondary importance. The effect of anisotropy for diffuse reflection $A'=0.90$ is greatest with higher values for clear-day conditions, practically absent for average cloudiness, and inverse with lower values for an overcast sky.

Figs. 4.12 and 4.13 are concerned with the influence of clouds in cases (a) and (b) as defined above.

Finally, Fig. 4.14 shows the influence of the albedo, the *type* of reflectivity of the ground, and the anisotropy of the sky radiation on the global radiation reaching a cylinder parallel to the earth's axis. The shading of the ground is neglected again.

These few examples will, it is hoped, stimulate interest in applied solar-radiation research with the view to clarifying the dependence of the intensity on field conditions. The various empirical and theoretical formulae now available may turn out to lead to similar conclusions as given already by many specialists. However, the problem is distinguished by the fact that the quantities investigated are functions of a large number of variables and this complicates the situation for the case of radiation falling on a horizontal plane, but allows for the first time valuable estimates of the radiation falling on tilted planes and cylindric, spheric or cubic receivers.

Diagrams 4.1–4.4 in the appendix (p. 148) are intended for practical use. For low latitudes only Tables 4.17–4.19 will allow estimates of daily sums of global radiation as a function of atmospheric turbidity. For higher latitudes Eq. (4.28 bis) must be applied.

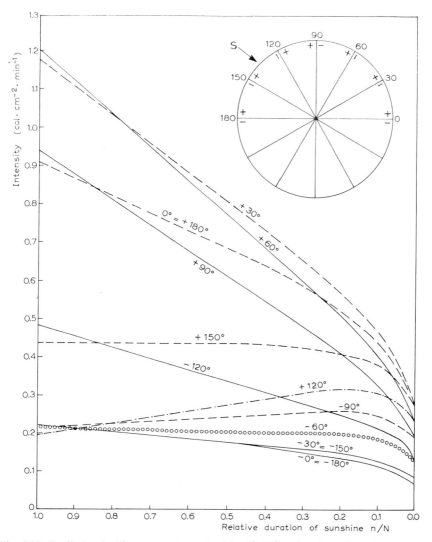

Fig. 4.12. Radiation incident on a plane tilted by 0°, 30°, 60°, 90°, 120°, 150° and 180° facing the solar vertical, or by −0°, −30°, −60°, −90°, −120°, −150° and −180° facing the ground, as a function of cloudiness (equator, equinox, cloudy days). Solar elevation 45°, albedo of the ground $A' = 0.25$.

References pp. 159–160

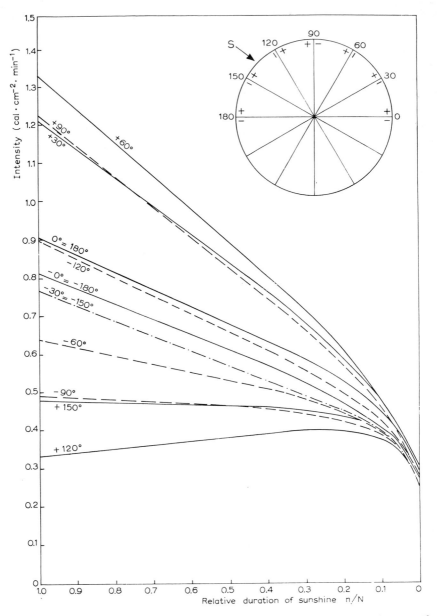

Fig. 4.13. Radiation incident on a plane tilted by 0°, 30°, 60°, 90°, 120°, 150° and 180° facing the solar vertical, or by −0°, −30°, −60°, −90°, −120°, −150° and −180° facing the ground, as a function of cloudiness (equator, equinox, cloudy days). Solar elevation 45°, albedo of the ground $A' = 0.90$.

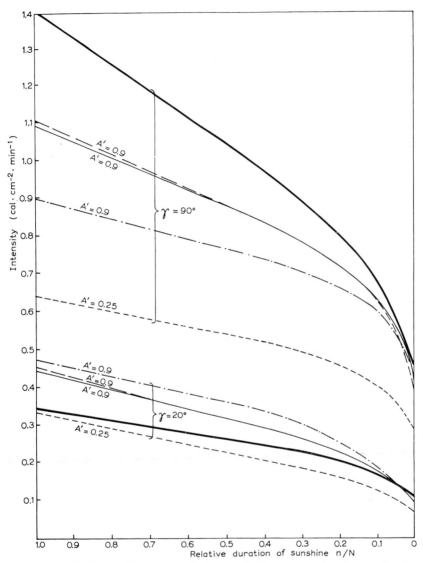

Fig. 4.14. Radiation incident on a cylinder parallel to the earth's axis as a function of cloudiness and albedo A' of the ground (equator, equinox, cloudy days), solar elevation 20° and 90°.

━━━ Horizontal plane
– – – Anisotropic sky
–·–·–· Specular reflecting ground
······ Diffuse reflecting ground, $A' = 0.25$
——— Diffuse reflecting ground, $A' = 0.90$

8. APPENDIX

Diagram 4.1. Solar elevation and related units

This diagram is based on the formula

$$\sin \gamma = \sin \delta \sin \varphi + \cos \delta \cos \varphi \cos t$$

where γ is the solar elevation, t is the hour angle, δ the solar declination, and φ the latitude of the location.

For a given location φ and solar declination δ, the daily variation of γ with t is plotted in the diagram. To obtain straight lines, which will facilitate interpolation, the plots are made using the $\sin \gamma$ and $\cos t$ scales, bearing in mind that:

$\sin \gamma = \sin \delta \sin \varphi$ for $\cos t = 0$ $(t = 90°;\ 06\ h)$
$\sin \gamma = \sin \delta \sin \varphi + \cos \delta \cos \varphi$ for $\cos t = 1$ $(t = 0°;\ 12\ h)$

In general, plots between these two positions are sufficient for values of δ between $-24°$ and $+24°$, taking δ to vary in steps of $2°$. For high-latitude stations the lines should be extended to $\cos t = -1.0$ and $\sin \gamma = -0.4$, plotting in steps of $1°$.

If the same diagram is to be used to evaluate data for different locations, additional plots must be constructed. These may be either drawn on transparent paper laid over the diagram, or a thread fixed by two pins at $\cos t = 0$ and $\cos t = 1$ may be preferred.

Additional true solar time (local apparent time) and solar elevation γ scales are provided below and on the right of the main field.

In other calculations the relative air mass m_r or the absolute air mass m must be known. These quantities are related by $m = (p/1000) \cdot m_r$, where p is the pressure in millibars.

To allow conversion for different values of m, use is made of the m-scales on the right-hand side of the diagram.

For a given elevation of the sun γ, m_r is easily found by extending a horizontal line from the $\sin \gamma$ (or γ)-axis to the m_r-axis. Extension of the same horizontal line to the m-axes corresponding to various pressures gives the corresponding m-values. The latter are read off interpolating between the sloping lines. A red vertical line indicating the mean pressure of the location may facilitate computation.

Diagram 4.1. Solar elevation and related units (a large-scale diagram can be found at the back of the book).   →

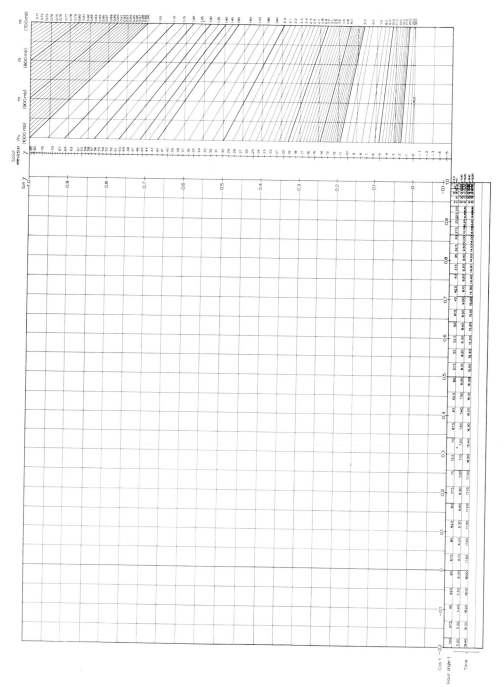

Diagram 4.2. Direct solar radiation as a function of atmospheric turbidity

This diagram gives the intensity of the direct solar radiation I_γ on the ground as a function of the solar elevation γ (or air mass m_r), the atmospheric water-vapour content w (in cm of precipitable water), and the turbidity coefficient B.

Since a pure atmosphere particularly reduces blue and ultraviolet radiation, and the turbidity acts principally on the visible radiation, and the absorption by water vapour acts almost exclusively on the infrared radiation, it is very difficult to construct a graph in which all these three elements are considered.

It was found necessary to start with mean conditions, *i.e.* $w = 2$ cm of precipitable water, and to draw the intensity I_γ of direct solar radiation on a logarithmic scale as the ordinate, and air mass m_r as the abscissa. Lines of constant B were then plotted (in field Y) in steps of 0.020, in the range $0 < B < 1$. This can be done with an accuracy of 1 % of the calculated values over the whole field. It must be emphasized that for a high turbidity even small deviations from the extinction law adopted ($\alpha = 1.5$ throughout the whole solar spectrum) may cause a considerable change in the resulting total direct radiation.

The second step is to draw graphs for the difference between I_γ values for $w = 2$ cm and for any other value of w. This has been done in the side-diagrams, for $w < 2$ cm on the left and $w > 2$ cm on the right. It is true that some residual differences occur, which may be considered in an additional table containing corrections between -4% and $+13\%$ as a function of w and m_r. Any other combination of diagrams would have resulted in greater final corrections, depending not only on w and m_r, but also on B.

Two scales are indicated at the bottom of the diagram. One of these gives m_r, the optical air mass corresponding to Bemporad's tables for sea level (but which can also be applied for stations at other levels by carrying out certain transformations; see below). The other scale is strictly limited to sea-level stations and gives the solar elevation γ. It is clear that the influence of solar elevation on solar intensity is large for low γ and vanishes for high γ.

The diagram is not completely self-sufficient. Four kinds of corrections have to be applied: (1) for atmospheric pressure, (2) for limitations in the construction of the graph, (3) for differences in the ozone content, and (4) for the reduction to mean solar distance.

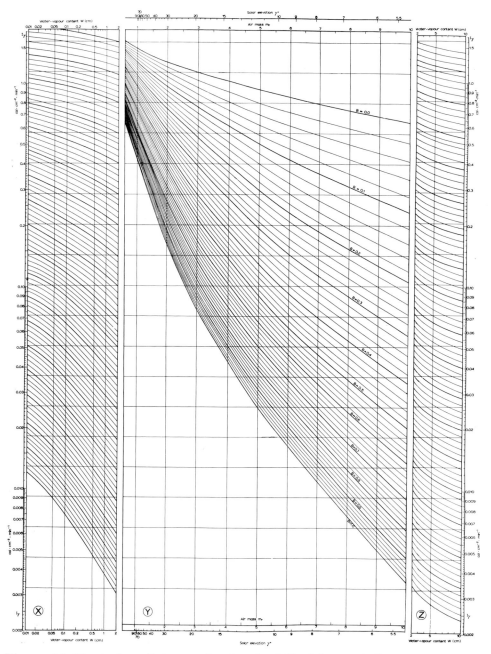

Diagram 4.2. Direct solar radiation as a function of turbidity (a large-scale diagram can be found at the back of the book).

(1) For atmospheric pressure $p \neq 1000$ mb (especially for stations not at sea level), calculate

$$m = 0.001 \cdot p \cdot m_r$$
$$B' = 1000 \cdot B/p$$
$$w' = 1000 \cdot w/p$$

With these three simple transformations as a function of p only the effect of altitude need be considered in the graph. (The vertical gradient of w and B as stated in Tables 4.3 and 4.4 is naturally not included in the above-mentioned formulae for B' and w'.)

(2) The residual error in the graph may be regarded as at the most $\pm 3\%$ if the I_γ found from the graph is multiplied by the factor given in Table 4.14.

TABLE 4.14. Correction factors for residual errors in Diagram 4.2.

w (cm)	m_r									
	1	2	3	4	5	6	7	8	9	10
0.01	0.96	0.98	1.03	1.04	1.05	1.07	1.09	1.10	1.12	1.13
0.02	0.97	0.99	1.03	1.04	1.05	1.06	1.07	1.09	1.11	1.12
0.04	0.97	1.00	1.02	1.03	1.05	1.05	1.06	1.08	1.10	1.11
0.10	0.98	1.00	1.02	1.03	1.04	1.04	1.04	1.06	1.08	1.09
0.20	0.99	1.00	1.01	1.02	1.03	1.03	1.03	1.04	1.05	1.06
0.40	0.99	1.00	1.01	1.01	1.02	1.02	1.02	1.02	1.03	1.04
1.00	1.00	1.00	1.00	1.00	1.01	1.01	1.01	1.01	1.02	1.02
2.00	1.00	1.00	1.00	1.00	1.00	1.00	1.00	1.00	1.00	1.00
4.00	1.00	1.00	0.99	0.99	0.99	0.99	0.99	0.99	0.99	0.99
10.00	1.00	0.99	0.98	0.98	0.98	0.98	0.98	0.98	0.99	0.99

(3) The whole diagram has been constructed for a mean atmospheric ozone content of $O_3 = 0.34$ cm NTP (IGY scale). If O_3 is considerably different, a correction as indicated in Table 4.10a or 4.10b has to be applied.

(4) For the reduction to mean solar distance, cf. Chapter 2.

Taking into account the four above-mentioned corrections it is possible to obtain the following information (for mean solar distance) on turbidity and solar radiation.

(a) *To find I_γ when γ, w and B are known*

Step 1: Enter field Y at the given value of γ or m_r, and move up to the curve corresponding to the given value of B.

Step 2: Move horizontally from this point to the boundary between X and Y if $w < 2$, and to the boundary between Y and Z if $w > 2$.

Step 3: Follow the curve (in field X or Z) until it intersects the given value of w. Interpolation between curves may be necessary.

Step 4: Read off I_γ horizontally opposite that point.

(*b*) *To find B when I_γ, γ and w are known*

This is essentially a procedure reverse to that described under (a).

Step 1: Enter field X (for $w < 2$) or field Z (for $w > 2$) at the given value of I_γ, and move horizontally to the given value of w.

Step 2: Follow the curve within field X or Z (this may have to be interpolated), to the boundary between this field and field Y.

Step 3: Enter field Y from the point reached on one of the boundaries in step 2, extending the line horizontally until it intersects the given value of γ.

Step 4: Read off B at this point, interpolating if necessary.

(*c*) *To find w when I_γ, γ and B are known*

Step 1: Enter field Y at the given value of γ (or m_r) and move up to the curve corresponding to the given B.

Step 2: Move horizontally to the boundaries between X and Y, and Y and Z, marking the points of intersection of the horizontal line with the above boundaries. For the next step, move to either field X or Z: if the given I_γ is above the marked points, use field X, if the given I_γ is below the marked points, use field Z.

Step 3: Follow the curve (into field X or Z) from the point on the boundary obtained in step 2 until it reaches the horizontal line through the given value of I_γ. This point corresponds to the required value of w.

From the central field Y, it is clear that the effect of turbidity has a gradient diminishing with air mass m_r and with turbidity B; this is due to the fact that the extinction of the shorter wavelengths gradually becomes complete, so that the remaining region is finally limited to the infrared, where the influence of turbidity is small. From the left-hand side diagram (field X) it appears that the effect of absorption by water vapour becomes more and more important at higher turbidities, because the absorption is limited to the infrared part of the solar spectrum, which gradually represents a more and more important fraction of the whole radiation. This does not appear so clearly from the right-hand side diagram (field Z); with a high atmospheric water content the far-infrared radiation is also completely absorbed, thus limiting the radiation to the central part of the spectrum, where the influences of both w and B are combined.

Diagram 4.3. Diffuse sky radiation above ground having albedo A=0.25 as a function of air mass m_r and turbidity B

This diagram gives the diffuse sky radiation D for ground albedo $A=0.25$, for a range of turbidity coefficients B, as a function of the air mass m_r. The curves are constructed for a precipitable water content $w=2$ cm and an ozone content of 0.35 cm NTP (IGY scale).

For high turbidity D approaches a 'saturation level'. Observations show that for a low sun this level is in practice reached even earlier than indicated by the diagram, and that for strong haze the sky radiation may even be reduced by some kind of absorption (cf. Robinson [45]). On the other hand, strong haze is often combined with rather large dust particles, in which case, as indicated on p. 118, sky radiation may be greater than indicated on the graph (cf. Valko [46]). Nevertheless, the graph gives a good idea of the sky radiation at sea level.

Three kinds of corrections should be considered: the influence of values of $w \neq 2$ cm, the influence of altitude, and the influence of solar distance (cf. Chapter 2).

For w values other than 2 cm, the diffuse sky radiation D should be corrected according to Table 4.15.

To allow for altitudes other than sea level, B must be replaced by

$$B' = B_{\text{sea level}} - 0.06 \cdot (1 - p/1000)$$

The influence of the ground albedo is very strong and is given by Diagram 4.4.

TABLE 4.15. Corrections (in mcal $\cdot$ cm^{-2} $\cdot$ min^{-1}) for D-values in Diagram 4.3 when $w \neq 2$ cm.

B	w (cm)							
	0.01	0.03	0.1	0.3	1.0	2.0	5.0	10.0
0.00	+3	+3	+2	+2	+1	0	−1	−2
0.10	+6	+5	+4	+3	+2	0	−2	−4
0.20	+7	+6	+4	+3	+2	0	−4	−8
0.40	+8	+7	+5	+4	+3	0	−6	−12

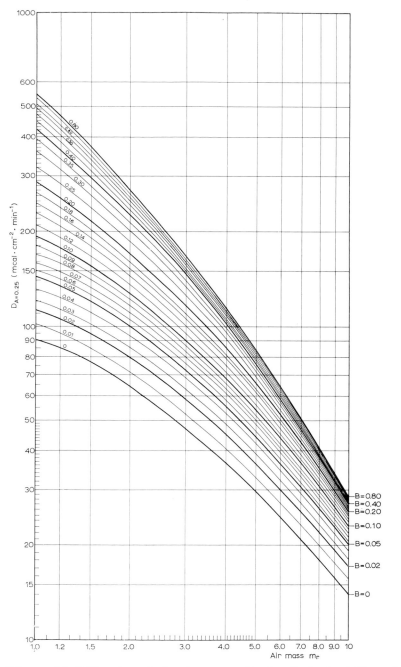

Diagram 4.3. Diffuse sky radiation above ground having albedo $A = 0.25$ as a function of air mass m_r and turbidity B.

Diagram 4.4. Diffuse sky radiation above ground with albedo $A \neq 0.25$

This diagram is based on Eqs. (4.11), (4.12), and (4.15), for a water-vapour content $w = 2$ cm and an ozone content of 0.35 cm NTP (IGY scale). Its object is to give a factor for converting $D_{\gamma, A=0.25}$ to the corresponding value of D for any other albedo $D_{\gamma, A}$.

This is done in the following manner:

Step 1: Draw a horizontal line (in the upper half of the diagram) through the appropriate value of B, and a line between the appropriate inclined lines corresponding to the given value of m_r. These intersect at point X in the upper part of the diagram.

Step 2: A perpendicular is now dropped from X to the curve (in the lower part of the diagram) plotted for the required value of A. Interpolation

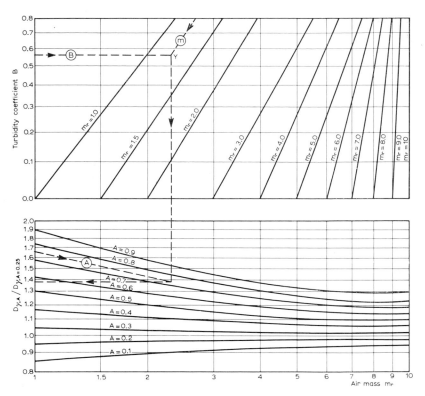

Diagram 4.4. Diffuse sky radiation above ground with albedo $A \neq 0.25$.

TABLE 4.16. Correction factors $f(w)$ for Diagram 4.4 when $w \neq 2$ cm.

w (cm)	0.01	0.03	0.1	0.3	1.0	2.0	5.0	10.0
$f(w)$	1.17	1.14	1.10	1.06	1.02	1.00	0.91	0.81

between lines of constant A may be necessary. The conversion factor $D_{\gamma,A}/D_{\gamma,A=0.25}$ is then read off horizontally on the left-hand ordinate.

This value is correct for $w = 2$ cm. For w values other than 2 cm, the following equation must be used:

$$D_{\gamma,A,w}/D_{\gamma,A=0.25,w=2} = 1 + (D_{\gamma,A}/D_{\gamma,A=0.25} - 1)f(w)$$

where $f(w)$ is a correction factor obtained from Table 4.16. Thus water vapour reduces the influence of A on the ratio $D_{\gamma,A}/D_{\gamma,A=0.25}$.

For very high sun, high albedo, and small turbidity, the contribution of multiple reflections may be higher than the primary diffusion reaching the ground, owing to the fact that the mean angle of incidence of diffuse reflected radiation is smaller than the elevation of the sun. For this reason reflected radiation is scattered along a greater path than direct radiation. In practice such extreme conditions will never occur because the albedo of snow is generally much lower than 0.9 when the sun is high.

Thus the use of Diagram 4.3 corrected for $w \neq 2$ cm, $p \neq 1000$ mb and solar distance, combined with Diagram 4.4 corrected for $w \neq 2$ cm will allow a fairly good estimate of diffuse sky radiation D_γ.

On the other hand, Diagram 4.2 gives the direct solar radiation I_γ as a function of the same parameters, providing corrections for $w \neq 2$ cm, $p \neq 1000$ mb, and the solar distance.

With both these values entered in Eq. (4.28 bis), radiation incident on any inclined plane at any point of the earth and at any time of the day may be calculated. For specific problems Eq. (4.28 bis) may easily be simplified. Anisotropy of the skylight is disregarded in Eq. (4.28 bis) but may be considered by Eq. (4.32).

Tables 4.17–4.19. Daily sums of global radiation G at low latitudes

For stations at low latitudes ($25°N > \varphi < 25°S$) at sea level (1000 mb) Tables 4.17, 4.18 and 4.19 allow estimates of daily sums of global radiation as a function of humidity w and turbidity B.

TABLE 4.17. Daily sums of global radiation ($w = 2$ cm, $p = 1000$ mb, $O_3 = 0.34$ cm, $B = 0$) in cal · cm^{-2} · day^{-1}.

Latitude	Month											
	Jan.	Feb.	Mar.	Apr.	May	June	July	Aug.	Sept.	Oct.	Nov.	Dec.
25°N	421	506	599	684	728	737	730	698	630	534	451	394
20°N	475	550	628	692	720	720	716	697	651	574	501	452
15°N	528	591	652	694	706	699	699	692	665	610	549	507
10°N	575	628	670	690	685	673	676	682	673	639	592	556
5°N	616	659	680	684	661	640	648	667	675	662	630	599
0°	654	684	687	672	631	605	615	648	675	680	664	643
5°S	690	704	688	655	599	564	581	625	665	692	693	681
10°S	722	717	684	631	561	522	540	597	653	697	715	718
15°S	747	725	673	603	516	476	492	563	634	699	734	745
20°S	765	728	658	566	470	424	445	524	609	696	748	769
25°S	779	726	634	523	420	368	392	482	580	686	757	787

TABLE 4.18. Correction factor F_w for $w \neq 2$ cm for application to Table 4.17.

w (cm)	F_w	w (cm)	F_w	w (cm)	F_w	w (cm)	F_w
0.25	1.10	0.82	1.05	2.3	0.99	4.6	0.94
0.30	1.09	1.00	1.04	2.7	0.98	5.2	0.93
0.43	1.08	1.20	1.03	3.0	0.97	6.0	0.92
0.55	1.07	1.40	1.02	3.5	0.96	6.7	0.91
0.67	1.06	1.70	1.01	4.0	0.95	7.6	0.90

TABLE 4.19. Correction factor F_B for $B > 0$ for application to Table 4.17.
F_B is a function of latitude φ and solar declination δ. The higher $\varphi - \delta$ the stronger the influence of turbidity on $G_{\gamma,B}$.

B	$\varphi - \delta$						
	0°	10°	20°	30°	40°	45°	50°
0.00	1.000	1.000	1.000	1.000	1.000	1.000	1.000
0.10	0.908	0.906	0.902	0.897	0.888	0.882	0.875
0.20	0.843	0.840	0.833	0.823	0.809	0.801	0.790
0.30	0.788	0.785	0.776	0.764	0.746	0.735	0.721
0.40	0.748	0.744	0.734	0.720	0.697	0.686	0.672
0.50	0.716	0.712	0.702	0.684	0.658	0.646	0.632
0.60	0.688	0.684	0.675	0.657	0.626	0.613	0.598
0.70	0.665	0.661	0.648	0.630	0.600	0.586	0.570

The rough value from Table 4.17 must be multiplied by the first factor F_w for precipitable water in the atmosphere and the second factor F_B for turbidity:

$$G = (Table\ 4.17) \cdot F_w \cdot F_B$$

Table 4.17 is drawn up for the 15th of each month, taking atmospheric pressure $p = 1000$ mb (Rayleigh atmosphere), turbidity $B = 0$, precipitable water $w = 2$ cm, ozone content $O_3 = 0.34$ cm and true solar distance.

These values may be corrected for actual precipitable water 0.25 cm $<$ $w < 7.6$ cm by the factor F_w given in Table 4.18.

The values may be corrected further for turbidity $0 < B < 0.70$ by the factor F_B given in Table 4.19 (calculated for the exponent α in Ångström's formula as 1.5). Because daily mean solar height (resp. solar declination δ) varies in the course of the year and with latitude φ the table is drawn up for values of $\varphi - \delta$ between 0° and 50°.

At higher latitudes and lower atmospheric pressure G must be estimated with the aid of Diagrams 4.1, 4.2 and 4.4 vor hourly intervals and then integrated over the whole day. For a given place (φ, p = const.) tables may also easily be drawn up.

REFERENCES

1. A. ÅNGSTRÖM, *Geografiska Annaler*, 11 (1929) 156.
2. J. W. STRUTT (LORD RAYLEIGH), *Scientific Papers*, 1, No. 9 (1881–1887) 10, 74.
3. F. W. P. GOETZ AND J. CABANNES, *Verhandl. Schweiz. Naturforsch. Ges.*, 120 (1940) 122.
4. F. E. FOWLE, *Astrophys. J.*, 35 (1912) 149; *Smithsonian Inst. Misc. Collections*, 68, No. 8 (1917).
5. W. MÖRIKOFER AND W. SCHÜEPP, *Arch. Meteorol. Geophys. Bioklimatol. Ser. B*, 2 (1951) 397.
6. W. L. GODSON, *Quart. J. Roy. Meteorol. Soc.*, 86 (1960) 301; H. WEXLER, W. B. MORELAND AND W. S. WEYANT, *Monthly Weather Rev.*, 88 (1960) 43; F. W. P. GOETZ, *Vierteljahresschr. Naturforsch. Ges. Zürich*, 89 (1944) 250.
7. IGY Instruction Manual of Ozone Measurements, *Annals of IGY*, V (1957), Pergamon, London.
8. J. HANN, *Lehrbuch der Meteorologie*, Tauchnitz, Leipzig (1901), p. 225.
9. H. FLOHN, *Z. Meteorol.*, 5 (1951) 148.
10. P. VALKO, *Untersuchung über die vertikale Trübungsschichtung der Atmosphäre*, Thesis, Basel (1960).
11. A. L. SHANDS, *U.S. Weather Bureau Tech. Paper*, No. 10 (1949).
12. G. MIE, *Ann. Physik*, 25 (1908) 377.

13. D. DEIRMENDJIAN AND Z. SEKERA, *J. Opt. Soc. Am.*, 46 (1956) 565; *Arch. Meteorol. Geophys. Bioklimatol. Ser. B*, 6 (1955) 452; *Ann. Geophys.*, 13 (1957) 286; 15 (1959) 218.

14. Z. SEKERA AND E. V. ASHBURN, Tables relating to Rayleigh Scattering of Light in the Atmosphere, U.S. Naval-Ordnance Test-Station, China Lake, Calif., *Navord Rept.*, No. 2061.

15. F. ALBRECHT, *Arch. Meteorol. Geophys. Bioklimatol. Ser. B*, 3 (1951) 220.

16. A. ÅNGSTRÖM, *Tek. Tidskr.*, (1924) 32.

17. Z. SEKERA AND D. DEIRMENDJIAN, *Tellus*, 6 (1954) 382.

17a. P. BENER, Investigation on the Spectral Intensity of Ultraviolet Sky and Sun + Sky Radiation, Davos, *Contract AF 61 (052)–54* (1960).

18. J. K. PAGE, Rept. First Intern. Soc. Bioclimatol. Vienna 1957, *Intern. J. Bioclimatol. Biometeorol.*, 2, Pt. IV (1958), Sect. D 2.

19. R. HOPKINSON, *Gt. Brit.*, *Dept. Sci. Ind. Res.*, *Bldg. Res.*, (1953) 268.

20. G. I. POKROWSKI, *Phys. Z.*, 30 (1929) 697.

21. O. ECKEL AND F. SAUBERER, *Meteorol. Z.*, 55 (1938) 151.

22. S. CHANDRASEKHAR, *Radiative Transfer*, Clarendon Press, Oxford (1950).

23. F. VOLZ, *Ber. Deut. Wetterdienstes*, 2, No. 13 (1954).

24. K. YA. KONDRATYEV AND M. P. MANOLOVA, *Solar Energy*, 4 (1960) 14.

25. P. MOON AND D. E. SPENCER, *Trans. Illum. Eng. Soc. (N. Y.)*, 37 (1942) 707.

26. F. TONNE AND W. NORMANN, *Z. Meteorol.*, 14 (1960) 166.

27. W. SCHÜEPP, *Arch. Meteorol. Geophys. Bioklimatol. Ser. B*, 10 (1960) 311.

28. M. ROBITZSCH, *Wetter*, 40 (1923) 28.

29. M. NICOLET AND R. DOGNIAUX, *Inst. Roy. Meteorol. Belg.*, *Mem.*, No. 40 (1950).

30. F. SAUBERER, *Wetter und Leben*, 8 (1956) 5–7.

31. M. BIDER, *Arch. Meteorol. Geophys. Bioklimatol. Ser. B*, 9 (1958) 199.

32. F. SAUBERER AND I. DIRMHIRN, *Arch. Meteorol. Geophys. Bioklimatol. Ser. B*, 6 (1954) 113.

33. A. ÅNGSTRÖM, *Monthly Weather Rev.*, 47 (1919) 797.

34. W. SCHÜEPP, *Météo Congo*, No. 7 (1953).

35. H. SHIELDRUP-PAULSEN, *Univ. Bergen Arbok Naturvitenskap. Rekke*, No. 7 (1948).

36. A. ÅNGSTRÖM, *Quart. J. Meteorol. Soc.*, 50 (1924) 121, 210; *Bioklimatische Beibl.*, 1 (1934) 6.

37. *WMO Guide to International Meteorological Instrument and Observing Practice*, WMO No. 8 T.P. 3 (1954), Chapter 9, revised June 1963.

38. F. W. P. GOETZ, *Klima und Wetter in Arosa*, Huber, Frauenfeld (1954).

39. H. HINZPETER, *Ber. Deut. Wetterdienstes U.S. Zone*, 42 (1952) 86; *Arch. Meteorol. Geophys. Bioklimatol. Ser. B*, 9 (1959) 60.

40. R. DOGNIAUX AND X. DE MARE D'AERTRYCKE, *Inst. Roy. Meteorol. Belg.*, *Publ. Ser. B*, No. 31 (1961).

41. *WMO–RA–1: Report of the 4th session WGR–RA–1 at Tunis, January 10–21, 1961*, Meteorol. Office, Tunis (1961).

42. D. Y. H. LIU AND R. C. JORDAN, *Solar Energy*, 4, No. 3 (1960) 1.

43. P. PETHERBRIDGE, *Quart. J. Roy. Meteorol. Soc.*, 81 (1955) 476.

44. S. FRITZ, *J. Opt. Soc. Am.*, 45 (1955) 820; *J. Meteorol.*, 11 (1954) 291; *Weather*, 12 (1957) 345.

45. G. D. ROBINSON, *Arch. Meteorol. Geophys. Bioklimatol. Ser. B*, 12 (1963) 19.

46. P. VALKO, Diffuse sky radiation as a function of meteorological and environmental factors, *Proc. 3rd Intern. Biometeorol. Congr.*, Pau 1963 (in the press); J. C. THAMS and P. VALKO, *Arch. Meteorol. Geophys. Bioklimatol. Ser. B* (in the press).

Chapter 5

THE ENERGY DISTRIBUTION IN THE SPECTRUM OF DIRECT AND SCATTERED RADIATION REACHING THE EARTH

Following the description in Chapter 1 of the solar spectrum outside the atmosphere, and the influence of atmospheric effects discussed in Chapter 3, the present Chapter will be devoted to the energy distribution in the spectrum of the direct radiation of the sun, the scattered radiation, and the total (sun + sky) radiation reaching the earth's surface. Theoretical and experimental results will be presented for the various regions of the spectrum.

1. THE ENERGY DISTRIBUTION OF DIRECT SOLAR RADIATION (I-RADIATION)

The first variable influencing the energy distribution in the solar spectrum which we will consider is the air mass m (cf. Chapter 3). To begin with, we shall assume that the atmosphere is perfectly clean and dry (Rayleigh atmosphere). The computed intensity for such an atmosphere is given in $mcal \cdot cm^{-2} \cdot min^{-1}$ in Table 5.1 for various values of the air mass. The last row of this Table gives the total intensity for each value of m.

Attenuation of the intensity in each spectral region is a consequence of Rayleigh scattering as explained in Chapter 3. It is expected that ultraviolet radiation will be reduced in intensity by scattering to a greater extent than the visible and infrared radiation, and this is in fact clearly shown by Table 5.1.

Fig. 5.1 shows computed values of the intensity as a function of the wavelength λ for $m = 0$, 1.1, 2.0, 3.0 and 5.0. The wavelength corresponding to the maximum energy is found to shift from about 4700 Å at $m = 0$ to about 6900 Å for a low sun.

According to Kalitin [1] the dependence of the energy distribution in the spectrum on the elevation of the sun $\gamma_\odot$ (or the air mass m) for the Ray-

TABLE 5.1. Intensity distribution (mcal · cm^{-2} · min^{-1}) for a Rayleigh atmosphere [2].

Wavelength (in μ)	Air mass m								
	0	0.5	1	2	3	4	6	8	10
0.28–0.30	2.6	1.3	0.7	0.2	—	—	—	—	—
0.30–0.32	11.5	7.0	4.2	1.6	0.6	0.2	—	—	—
0.32–0.34	22.8	15.5	10.6	4.9	2.3	1.1	0.2	—	—
0.34–0.36	23.7	17.6	13.1	7.2	4.0	2.2	0.7	0.2	0.0
0.36–0.38	30.5	24.0	19.0	11.8	7.4	4.6	1.8	0.7	0.3
0.38–0.40	40.0	33.2	27.4	18.8	12.9	8.8	4.2	2.0	0.9
0.40–0.42	55.0	47.2	40.4	29.7	21.7	16.0	8.7	4.7	2.6
0.42–0.44	57.0	50.2	44.3	34.4	26.8	20.8	12.5	7.6	4.7
0.44–0.46	61.0	54.9	49.5	40.2	32.6	26.6	17.6	11.6	7.6
0.46–0.48	62.9	57.7	52.8	44.3	37.1	31.2	22.0	15.5	10.9
0.48–0.50	62.5	58.0	53.8	46.6	40.3	35.4	26.0	20.1	14.2
0.50–0.52	59.7	56.0	52.7	46.5	41.1	36.2	28.2	22.0	17.1
0.52–0.54	57.3	54.4	51.5	46.2	41.5	37.3	30.0	24.3	19.7
0.54–0.56	55.5	53.0	50.7	46.2	42.2	38.6	32.2	26.8	22.4
0.56–0.58	54.6	52.6	50.4	46.6	43.1	39.9	34.0	29.1	24.8
0.58–0.60	54.3	52.5	50.7	47.3	44.2	41.4	36.3	31.6	27.4
0.60–0.62	52.8	51.2	49.7	46.8	44.0	41.5	36.6	32.6	28.5
0.62–0.64	50.3	49.0	47.7	45.2	42.8	40.7	36.6	32.9	29.5
0.64–0.66	48.4	47.2	46.1	44.1	42.0	40.1	36.6	33.3	30.5
0.66–0.70	92.3	90.7	88.8	85.4	82.2	79.0	73.1	67.2	62.6
0.70–0.74	83.1	81.8	80.6	78.1	75.7	73.4	69.0	65.0	61.4
0.74–0.80	106.4	105.0	103.8	101.5	99.1	97.1	92.6	88.7	84.0
0.80–0.90	143.4	142.4	141.0	138.8	136.5	134.3	130.4	126.1	122.5
0.90–1.00	113.4	112.8	112.2	111.1	110.0	108.8	106.5	105.9	102.5
1.00–2.00	426.0	425.0	424.6	423.0	421.6	419.7	417.1	413.9	411.0
>2.00	113.0	113.0	113.0	113.0	113.0	112.8	112.7	112.7	112.6
Total	1940.0	1853.4	1779.3	1659.5	1564.7	1487.7	1365.6	1274.5	1198.0

leigh atmosphere can be described by the family of curves shown in Fig. 5.2. In this figure the energy is plotted along the horizontal axis in cal · cm^{-2} · min^{-1} and the values of $\gamma_\odot$ are plotted along the vertical axis. The upper horizontal row of numbers shows the wavelengths in microns. The numbers in the squares indicate the amount of energy in the given spectral region as a percentage of the total energy. For example, the infrared accounts for 49.6% of the total energy when $\gamma_\odot = 60°$ and for 52.2% when $\gamma_\odot = 30°$. In order to determine the absolute energy in a given spectral region the numbers on the horizontal axis have to be subtracted. Thus for infrared with $\gamma_\odot = 60°$ the absolute energy is $1.60 - 0.80 = 0.80$ cal · cm^{-2} · min^{-1}.

A further demonstration of the dependence of the intensity of radia-

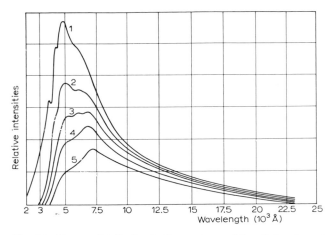

Fig. 5.1. Energy distribution in the spectrum of I-radiation for a dry and clean atmosphere, as a function of the air mass m [2].
(1) $m = 0$, (2) $m = 1.1$, (3) $m = 2.0$, (4) $m = 3.0$, (5) $m = 5.0$.

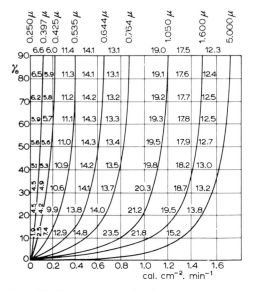

Fig. 5.2. Energy distribution in the spectrum of I-radiation as a function of $\gamma_{\odot}$ [1].

tion at different wavelengths on the air mass m is provided by the Bouguer–Lambert law, which in logarithmic form reads (cf. Chapter 3):

$$\log I_{m\lambda} = \log I_{0\lambda} + m \log q_\lambda \tag{5.1}$$

where $I_{0\lambda}$ is the intensity outside the atmosphere, $I_{m\lambda}$ is the intensity after traversing an air mass m, and q_λ is the transmissivity at the particular wavelength λ. When q_λ is independent of time, two points are sufficient for the determination of the dependence of $I_{m\lambda}$ on m at given wavelength λ (see Fig. 5.3 in which the slope of the straight lines is equal to $\log q_\lambda$).

When the radiation is not monochromatic, the straight lines become curved as shown in Fig. 5.3 in which curve 1 is for the total Rayleigh radiation, curve 2 is for the total solar radiation, curve 3 is for the red region and curve 4 is for radiation in the range of 3000–30,000 Å. Extrapolation to $m=0$ is not an easy matter because the form of the functions q_λ and $I_{0\lambda}$ is not well known. Different authors appear to have used different values for these functions [3,4].

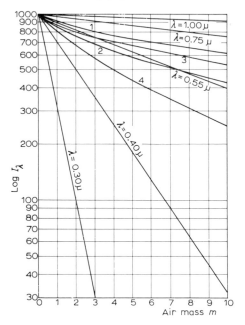

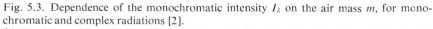

Fig. 5.3. Dependence of the monochromatic intensity I_λ on the air mass m, for monochromatic and complex radiations [2].
(1) Total Rayleigh radiation, (2) total solar radiation, (3) red region, (4) radiation in the range 3000–30,000 Å.

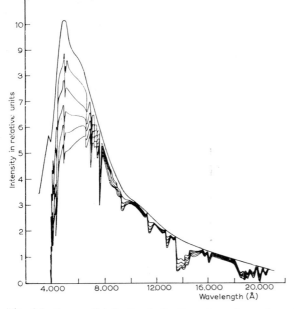

Fig. 5.4. Energy distribution in the I-spectrum for a clean atmosphere and various values of the air mass m [5].

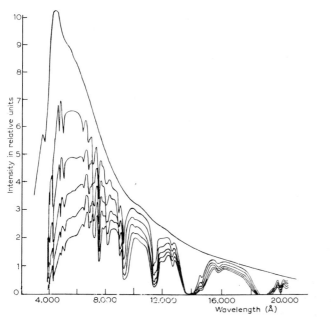

Fig. 5.5. Energy distribution in the I-spectrum for a turbid and humid atmosphere and various values of the air mass m [5].

The above results are for a perfectly clean and dry atmosphere, and in spite of the fact that such an atmosphere is very seldom found, the data are of great importance since they can be applied at all geographical points with the same $\gamma_\odot$ and h.

Moreover, they are the only results which are amenable to accurate and comparatively simple computations and can, therefore, serve as reference for other more complicated conditions in the atmosphere.

The combined effect of the air mass m, the water vapour content w and all other components such as dust, is very well described by the two sets of curves shown in Figs. 5.4 and 5.5.

The uppermost curve in each set is for $m=0$, whilst the others correspond to $m=1, 2, 3, 4$ and 5. The set shown in Fig. 5.4 represents results of measurements for a clear sky and an almost dry atmosphere at high altitudes, *i.e.* for a Rayleigh atmosphere. The curves of Fig. 5.5 were obtained at sea level, *i.e.* for a more humid and turbid atmosphere. These curves show the considerable effects of the water content and turbidity on radiation reaching the earth. All the curves of Fig. 5.5 are lower at the same m value than those in Fig. 5.4. The curves in Fig. 5.5 differ from each other more than those in Fig. 5.4. Particularly important are the minima in both sets. The strong minima at 7500, 9000, 11,500–12,000, 14,500 and 19,500 Å (Fig. 5.5) are due to water-vapour absorption, while the other minima, *e.g.* those at 14,650, 16,000 and 20,400 Å are due to the absorption by carbon dioxide.

2. THE ENERGY DISTRIBUTION OF SCATTERED SKY RADIATION (D-RADIATION)

The spectral curves discussed above were obtained in such a way as to exclude all scattered radiation. They represent the conditions on a plane perpendicular to the beam. In order to determine the spectral curves for scattered radiation, the instrument is usually directed towards various points on the sky at different angular distances ψ from the sun, and the various sky zones are examined with the sun itself obstructed. When the instrument is freely exposed on a horizontal plane to the entire hemisphere and only the sun is obstructed, a mixed spectral curve is obtained, corresponding to the entire scattered radiation for the particular point, *i.e.* to the particular astronomical and atmospheric conditions. Mixed spectral curves can be obtained for less than the total hemisphere by using an inclined plane. The energy distribution in the D-radiation spectrum depends on the ψ of the zone, the state of the atmosphere, the sun's elevation $\gamma_{\odot}$ and the degree of cloudiness.

It is to be expected from the discussion in Chapter 3 that the spectral composition of D-radiation will differ from that of the I-radiation, since the scattering is proportional to λ^{-4}.

The spectral curves of D-radiation can easily be computed for a Rayleigh atmosphere and a cloudless sky. They can also be calculated for certain distributions of water drops on the basis of the Mie–Debye theory with the refractive index $n = 1.33$. The results of such computations are shown in Table 5.2 for 'subozonal radiation' (below the ozone layer).

TABLE 5.2. Intensity in narrow spectral regions for various water drop distributions [6].

Wavelength (μ)	I_0-radiation (mcal $\cdot$ cm^{-2} $\cdot$ min^{-1})	D-radiation (mμcal $\cdot$ cm^{-2} $\cdot$ min^{-1}) Number of water drops per cm^3 of clear air			
		0	100 ($r = 0.1\,\mu$)	25 ($r = 0.5\,\mu$)	5 ($r = 1.0\,\mu$)
0.28–0.30	2.6	4.4	0.05	0.14	1.0
0.30–0.32	11.5	14.4	0.27	0.78	4.6
0.32–0.34	21.8	21.9	0.44	1.62	9.2
0.34–0.36	31.3	23.5	0.54	2.78	12.8
0.36–0.38	35.2	20.8	0.51	3.52	14.6
0.38–0.40	36.0	16.9	0.45	3.78	15.1
0.42–0.44	54.3	17.2	0.52	6.41	23.2
0.46–0.48	62.6	13.7	0.46	7.65	25.6
0.50–0.52	59.7	9.3	0.36	7.35	20.6
0.56–0.58	54.6	4.4	0.25	6.67	14.4
0.64–0.66	48.4	2.8	0.136	5.48	18.6
0.70–0.72	42.9	1.8	0.090	4.29	20.0
0.78–0.80	33.3	0.8	0.046	3.04	18.0
0.86–0.88	27.1	0.5	0.027	2.11	15.8
0.98–1.00	21.0	0.2	0.015	1.36	12.4

The first column of Table 5.2 gives the spectral ranges, and the second column the intensity of I-radiation. The remaining columns give the intensity of the D-radiation for various drop distributions. As can be seen, the number and the dimensions of the drops exert a considerable effect on the intensity. One feature of these intensity distributions should be mentioned, namely the fact that the absolute maximum for a Rayleigh atmosphere and for the same volume containing 5 drops of water per cm^3 (without air) with

$r=1\,\mu$ occurs at the same wavelength. The spectral curves differ, however, the second curve having a relative maximum at $\lambda=0.71\,\mu$.

In order to calculate the energy distribution in the spectrum due to scattering by pure air containing a given number of water drops of known radii, the intensities of the scattered radiation due to the Rayleigh atmosphere and to the water-drop distribution have to be added in each spectral region.

3. SPECTRA OF VARIOUS ZONES OF THE SKY AND OF THE ENTIRE HEMISPHERE

One example of such spectra is given in Table 5.3 for a plane facing the zone in the sun's vertical and in the opposite vertical. The radiation intensity in the sun's vertical is higher and the maximum intensity of the scattered radiation occurs near the horizon, the intensity reaching a maximum when the sun is at the zenith.

TABLE 5.3. Intensity in various spectral regions and in the total sky radiation in a unit spatial angle in mcal $\cdot$ cm^{-2} $\cdot$ min^{-1} for various ψ values [6].

	Sun's vertical					Opposite vertical				
	85°	65°	45°	25°	5°	5°	25°	45°	65°	85°
$\psi = 0°$										
< 0.4 μ	5.0	5.2	5.1	5.3	5.4	5.4	5.3	5.1	5.2	5.0
0.4–0.6 μ	25.3	11.7	9.6	9.4	9.4	9.4	9.4	9.6	11.7	25.3
> 0.6 μ	14.0	4.5	3.5	3.4	3.4	3.4	3.4	3.5	4.5	14.0
Total	44.4	21.4	18.3	18.0	18.3	18.3	18.0	18.3	21.4	44.4
$\psi = 30°$										
< 0.4 μ	6.2	7.0	6.3	5.6	4.8	4.4	3.7	3.6	4.2	5.5
0.4–0.6 μ	32.6	16.4	12.3	10.1	8.5	7.8	6.7	6.8	9.9	28.9
> 0.6 μ	18.3	6.4	4.5	3.6	3.1	2.8	2.5	2.5	3.8	16.2
Total	57.0	29.7	23.1	19.4	16.3	15.0	12.9	12.8	17.9	50.6
$\psi = 60°$										
< 0.4 μ	5.6	6.4	4.9	3.7	2.7	2.4	2.2	2.7	4.2	5.2
0.4–0.6 μ	41.3	18.2	11.4	7.9	5.8	5.1	4.8	6.3	12.2	37.9
> 0.6 μ	24.6	7.4	4.4	3.0	2.2	1.9	1.8	2.4	5.0	22.6
Total	71.6	32.1	20.8	14.6	10.7	9.5	8.8	11.5	21.4	65.7
$\psi = 90°$										
< 0.4 μ	0.07	0.68	0.56	0.40	0.32	0.32	0.38	0.52	0.65	0.06
0.4–0.6 μ	14.6	6.9	3.8	2.4	1.9	1.8	2.2	3.5	6.6	14.4
> 0.6 μ	19.0	5.0	2.6	1.6	1.2	1.2	1.5	2.4	4.8	18.9
Total	33.7	12.6	6.9	4.4	3.4	3.4	4.1	6.5	12.0	33.4

TABLE 5.4. Intensity on a unit horizontal area in mcal $\cdot$ cm^{-2} $\cdot$ min^{-1} from sky zones of 10° width [6].

	Sky zone									Total hemi-sphere
	5°	15°	25°	35°	45°	55°	65°	75°	85°	
$\psi = 0°$										
$< 0.4\,\mu$	0.5	1.5	2.2	2.7	2.8	2.6	2.2	1.5	0.9	16.5
0.4–$0.6\,\mu$	0.9	2.6	3.9	4.8	5.3	5.3	4.9	4.2	2.4	34.3
$> 0.6\,\mu$	0.3	0.9	1.4	1.8	1.9	2.0	1.9	1.7	1.3	13.3
Total	1.7	4.9	7.6	9.3	10.0	9.9	9.0	7.4	4.2	64.0
$\psi = 30°$										
$< 0.4\,\mu$	0.4	1.2	1.9	2.4	2.6	2.5	2.2	1.5	0.5	15.2
0.4–$0.6\,\mu$	0.8	2.2	3.5	4.4	5.0	5.2	5.1	4.5	2.6	33.4
$> 0.6\,\mu$	0.3	0.8	1.3	1.6	1.8	2.0	2.0	1.9	1.5	13.1
Total	1.5	4.3	6.7	8.4	9.4	9.7	9.2	7.9	4.6	61.8
$\psi = 60°$										
$< 0.4\,\mu$	0.2	0.7	1.2	1.6	1.8	1.9	1.8	1.3	0.4	11.0
0.4–$0.6\,\mu$	0.5	1.5	2.5	3.5	4.3	4.9	5.2	5.0	3.0	30.3
$> 0.6\,\mu$	0.2	0.6	1.0	1.3	1.6	1.9	2.1	2.2	1.8	12.7
Total	1.0	2.9	4.7	6.3	7.7	8.7	9.1	8.4	5.2	54.0
$\psi = 90°$										
$< 0.4\,\mu$	0.03	0.09	0.15	0.21	0.25	0.26	0.22	0.11	0.01	1.32
0.4–$0.6\,\mu$	0.18	0.53	0.90	1.29	1.67	2.01	2.18	2.05	1.04	11.87
$> 0.6\,\mu$	0.11	0.35	0.60	0.86	1.13	1.40	1.59	1.68	1.36	9.07
Total	0.32	0.97	1.65	2.36	3.06	3.66	4.00	3.84	2.40	22.26

For comparison, the scattered radiation reaching a horizontal plane from 10° zones is given in Table 5.4, in which the last column gives the radiation from the entire hemisphere.

Fig. 5.6 shows spectral curves for different zones at the same altitude

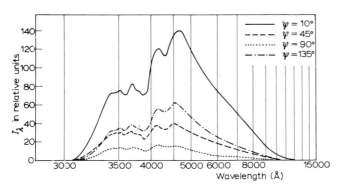

Fig. 5.6. Sky radiation for different values of ψ; $\gamma_\odot = 18°$ [7].

References pp. 194–195

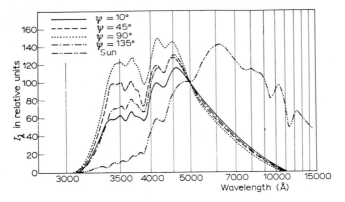

Fig. 5.7. Sky radiation for various values of ψ plotted relative to the energy at 5000 Å [7].

of the sun. The intensity is given in arbitrary units and the turbidity factor is $T = 1.4$. The maxima and minima of these curves are not characteristic of the scattered radiation but originate outside the atmosphere. It is clear from these curves that as the angular distance ψ varies from 90° to 10°, the intensity of the ultraviolet radiation changes by a factor of 5.6, while the intensity of visible radiation changes by a factor of 8.7.

Fig. 5.7 represents the same spectral curves in another way, showing the changes in the different spectral regions more clearly.

It is seen that the maximum at 4550 Å moves towards shorter wavelengths with increasing share of the ultraviolet radiation. The spectral curves

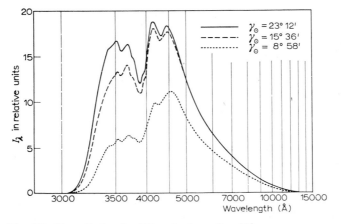

Fig. 5.8. Sky radiation for different values of $\gamma_\odot$ [7].

of D-radiation are found to be displaced towards shorter wavelengths as compared with the I-spectrum. In order to study the influence of $\gamma_\odot$ on the spectral composition of the scattered radiation, the curves must be constructed for constant values of ψ. Fig. 5.8 shows such curves for $\psi = 90°$ and three different values of $\gamma_\odot$ ($T = 1.4$).

The relative distribution (100 corresponding to a wavelength of 5000 Å) derived from these curves is shown in Fig. 5.9. It is clear that the short-wave part of the spectrum in general increases with $\gamma_\odot$ and the maximum at 4550 Å moves towards shorter wavelengths.

Fig. 5.10 shows the intensity of the scattered radiation for different values of the turbidity. It is clear that, for given $\gamma_\odot$ and ψ, the main effect of the increased turbidity is on the ultraviolet part of the radiation.

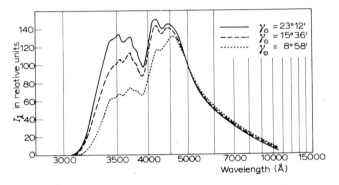

Fig. 5.9. Sky radiation for different values of $\gamma_\odot$ plotted relative to the energy at 5000 Å [7].

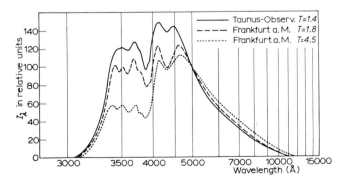

Fig. 5.10. Sky radiation for different values of the turbidity coefficient T, plotted relative to the energy at 5000 Å [7].

References pp. 194–195

The D-radiation intensity from the entire hemisphere on a horizontal plane is given in Table 5.5 for various values of $z_\odot$ in three spectral regions as a percentage of the total.

TABLE 5.5. The intensity in % of the total sky radiation in three spectral ranges as a function of $z_\odot$ [6].

$z_\odot$	Wavelength			
	$< 0.4\,\mu$	$0.4–0.6\,\mu$	$> 0.6\,\mu$	total
0°	25.8	53.5	20.7	100
30°	24.6	54.2	21.2	100
45°	23.2	54.8	22.0	100
60°	20.4	56.1	23.5	100
75°	14.6	58.2	27.2	100
87°	5.9	53.3	40.8	100

The proportion of ultraviolet radiation is found to decrease as $z_\odot$ increases from zero to 87° by a factor of approximately 5. On the other hand radiation above 6000 Å is found to increase by a factor of 2 only. It is interesting to note that the percentage of radiation bearing maximum energy (the central part of the spectrum) remains constant.

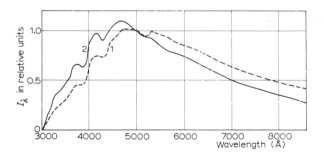

Fig. 5.11. Energy distribution plotted relative to the energy at 5000 Å [8].
(1) Combined solar and sky spectrum, (2) D-radiation.

Fig. 5.11 shows a spectral curve for the scattered radiation from the entire hemisphere on a horizontal plane (curve 2). The curve is plotted in relative units with the energy at 5000 Å taken as unity.

4. THE ENERGY DISTRIBUTION OF COMBINED SOLAR AND SKY RADIATION (G-RADIATION)

The relative spectral composition of this radiation between 3150 and 8000 Å is practically constant during the day and is slightly affected by the turbidity. Curve 1 in Fig. 5.11 shows a typical curve based on twelve observations. It is seen that this radiation contains a larger proportion of short-wave energy than the scattered radiation. However, the maximum occurs in the region of larger wavelengths. The effect of clouds on this radiation is illustrated in Fig. 5.12.

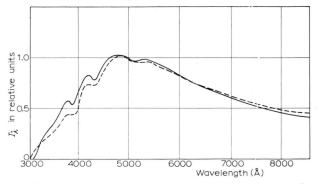

Fig. 5.12. Energy distribution relative to the energy at 5000 Å in the combined solar and sky spectrum [9].
Solid line: cirrus clouds; broken line: cumulus clouds.

For practical applications to climatology, biology, photosynthesis and direct utilization of solar energy, it is necessary to possess a detailed knowledge of the energy distribution in the solar spectrum. A knowledge of changes in this distribution which are due to astronomical and atmospheric effects is also necessary.

5. THE ULTRAVIOLET REGION 2800–3800 Å

Although this spectral region covers only a small fraction of the total solar energy reaching the earth (cf. Tables 1.2 and 5.1), it is however of considerable importance. Since the energy is given by $E = h\nu = 1.98 \times 10^{-8}/\lambda$ ergs, where h is Planck's constant, it is clear that the energy increases rapidly with decreasing wavelength λ.

TABLE 5.6. The biological effects of ultraviolet radiation [10].

Effect	Limit of activity
Bactericidal	3050 Å
Formation of vitamin D	3100 Å
Erythema	3150 Å
Albumen coagulation	3250 Å

The important biological effects of ultraviolet radiation are summarized in Table 5.6. All these effects are dependent on the wavelength as shown in Fig. 5.13.

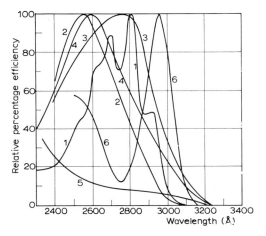

Fig. 5.13. Biological effects of ultraviolet radiation [11].
(1) Production of vitamin D, (2) bactericidal effects, (3) stoppage of tissue growth, (4) albumen coagulation, (5) haemolysis, (6) erythema.

Ultraviolet I-radiation

Fig. 5.14 shows the energy distribution in the ultraviolet I-spectrum at sea level. In addition to the decrease in the energy with increasing air mass m, these curves also show further important effects, namely the rapid decrease on the short wavelength side, and the fact that the wavelength at which the intensity becomes zero increases with the air mass. Both effects are a consequence of atmospheric scattering (cf. Chapter 3).

Fig. 5.15 shows the ultraviolet curves at $h = 3730$ m (Climax, Colorado). The curves in Figs. 5.14 and 5.15 are somewhat smoothed out because not

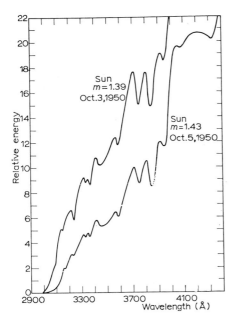

Fig. 5.14. Energy distribution in the ultraviolet I-spectrum at Washington, D.C., for $m = 1.39$ and 1.43 [12].

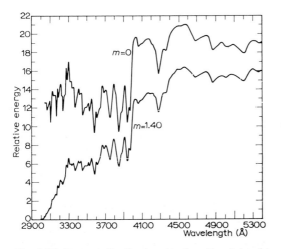

Fig. 5.15. Energy distribution in the ultraviolet I-spectrum at Climax, Colorado, for $m = 1.40$ and $m = 0$ (computed) [13,14].

all the Fraunhofer lines were detected by this instrument. The minimum wavelength at which the curves cut the wavelength axis depends on m but only slightly on h. This may be explained by the fact that the ozone layer in the upper atmosphere is responsible for a cut-off at short wavelengths.

An example of the variation of λ_{min} with the altitude h is given below [1].

h (metres)	50	116	1620	3136	4560
λ_{min} (Å)	2912.6	2912.4	2913.6	2911.0	2912.1

The dependence of λ_{min} on log sin $\gamma_\odot$ is linear. For example, the result for Pavlovsk (USSR) is (see Table 5.7):

$$\lambda_{min} = a - b \log \sin \gamma_\odot \tag{5.2}$$

where $a = 290$ mμ, $b = 25.8$ mμ. As a further example, Table 5.8 shows the variation of λ_{min} with $\gamma_\odot$ and the seasons in Switzerland. The minimum value of λ occurs in the autumn as a consequence of the change in the thickness of the ozone layer which is a minimum in the autumn.

TABLE 5.7. λ_{min} as a function of $\gamma_\odot$ [15].

$\gamma_\odot$	λ_{min} (Å)	$\gamma_\odot$	λ_{min} (Å)
1°	4200	20°	3040
2°	3820	25°	3020
3°	3520	30°	3000
5°	3270	35°	2980
7°	3180	40°	2970
10°	3120	45°	2960
15°	3060	50°	2950

TABLE 5.8. λ_{min} (in Å) as a function of $\gamma_\odot$ in various seasons [28].

Season	$\gamma_\odot$					
	10°	20°	30°	40°	50°	60°
Spring	3160	3080	3040	3020	2980	2980
Summer	3190	3100	3060	3010	2990	2980
Autumn	3120	3050	3020	3000	—	—
Winter	3140	3080	3070	—	—	—
Year (mean)	3160	3080	3040	3020	2990	2980

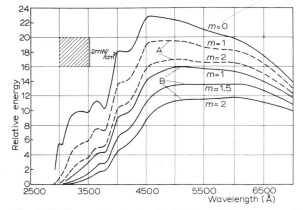

Fig. 5.16. Energy distribution in the ultraviolet I-spectrum [16].
(*A*) $h = 1789$ m, (**B**) $h = 0$.

The elevation above sea level has a very important effect on the ultraviolet radiation. This is illustrated in Fig. 5.16 in which curves A correspond to $m = 1$ and 2 at $h = 1789$ m (Mount Wilson) and curves B to $m = 1$, 1.5 and 2 at sea level (Washington, D.C.). It is evident from these curves that the amount of ultraviolet radiation increases with h at constant air mass. Fig. 5.17 shows the ultraviolet spectrum at various values of h.

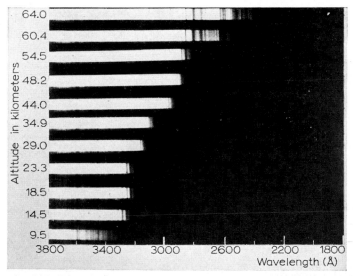

Fig. 5.17. Ultraviolet spectra of the sun at different values of h from 9.5 to 64 km [17].

Ultraviolet D-radiation

It has already been mentioned that ultraviolet radiation is strongly scattered because of the λ^{-4} law. Owing to the quantity and the non-directional properties of D-radiation, this radiation is very important in the ultraviolet. It is found in the optical shadow of buildings, trees and other objects and is of great biological importance because it is possible to use

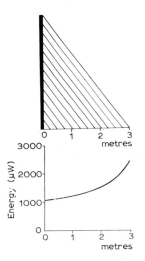

Fig. 5.18. Ultraviolet D-radiation inside the optical shadow [18].

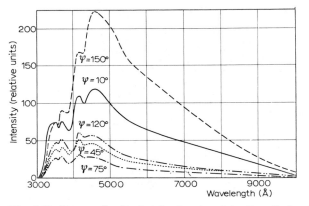

Fig. 5.19. Energy distribution in the ultraviolet D-radiation for a clear sky ($\gamma_{\odot} = 24°$) [19].

D-radiation to obtain quite large ultraviolet doses without direct exposure to the sun. Fig. 5.18 shows the penetration of this radiation into the shadow of a building.

The spectral curves of ultraviolet D-radiation are different from those of ultraviolet I-radiation. The total amount of ultraviolet D-radiation and its spectral composition depends on the particular zone of the sky. Fig. 5.19 shows some typical results for a very clear atmosphere and a cloudless sky, the energy at 5000 Å being taken as 100.

The results of Bener's experiments (see below) for the ultraviolet D-radiation are shown in Figs. 5.24–5.27 (pp. 183–185).

Ultraviolet G-radiation

Recently a very detailed experimental study—the first of its kind in extent and instrumentation—was carried out by Bener [29] at the Physical–Meteorological Observatory in Davos–Platz, Switzerland. The instrument used for this work is described in Chapter 7 (Fig. 7.53). Some of his results fitting into the frame of this chapter will be given here.

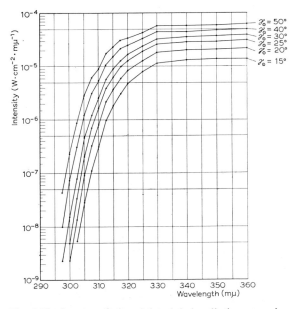

Fig. 5.20. Spectra of ultraviolet global radiation at various solar altitudes $\gamma_\odot$ and for an amount $X = 0.250$ cm of atmospheric ozone [29].

In Fig. 5.20, the dependence on λ of the energy in the spectrum of ultraviolet global radiation is given, the values of $\gamma_\odot$ ranging from 15° to 50°. This dependence is strongly expressed in the U.V.B. (the biologically active part of the ultraviolet spectrum), increasing with increasing λ. For the same $\Delta\lambda$ the increase of intensity depends on $\gamma_\odot$.

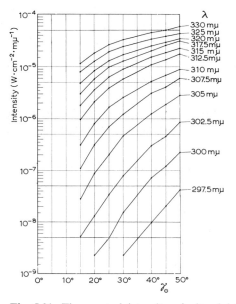

Fig. 5.21. The spectral intensity of ultraviolet global radiation. Dependence on solar altitude for various wavelengths and for an amount $X = 0.250$ cm of atmospheric ozone [29].

In Fig. 5.21, the same results are presented as a function of $\gamma_\odot$ for constant values of λ from 297.5 mμ to 330 mμ.

In both cases the ozone layer in the upper atmosphere was of the same thickness 0.250 cm, expressed in the scale used before the IGY. According to the new scale these values have to be multiplied by 1.33.

The very strong influence of the ozone layer thickness is shown in Fig. 5.22. This effect is seen to decrease with increasing λ for the same value.

Another factor affecting the energy distribution in the spectrum of the ultraviolet G-radiation is the season, all other factors being constant. This

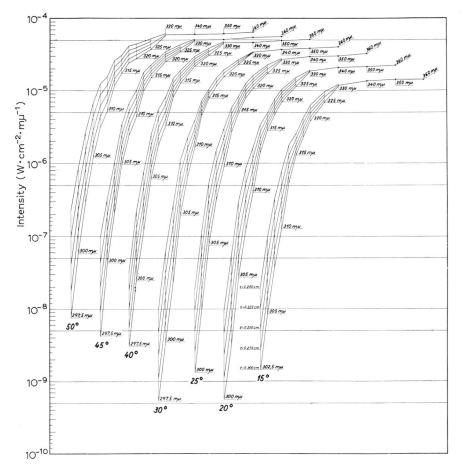

Fig. 5.22. Ultraviolet global spectra for various solar altitudes and various amounts X of atmospheric ozone [29].

effect is expressed by a relation $Q_{WS} = G_W/G_S$, where W stands for winter and S for summer, and is a consequence of the albedo of the earth, as caused by snow in winter. Whereas the theoretical relation is independent of $\gamma_\odot$ the measured values show a strong dependence on $\gamma_\odot$ (Fig. 5.23).

An attempt was made to measure the influence of turbidity (see Chapter 3) on the ultraviolet G-energy. No clear result was achieved.

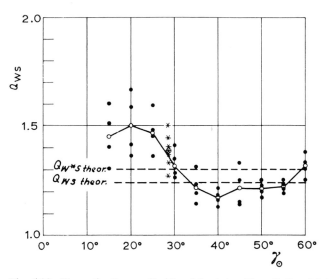

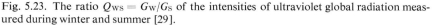

Fig. 5.23. The ratio $Q_{WS} = G_W/G_S$ of the intensities of ultraviolet global radiation meas-
ured during winter and summer [29].
● Figures relating to $\lambda = 330$–360 mμ and to the mean intensities measured during the
winter and summer periods.
○ Mean values, averaged over the interval 330 m$\mu \leqq \lambda \leqq 360$ mμ.
* Figures Q_{W*S} relating to the first snowfall in October 1958.
- - - - - Theoretical values.

Figs. 5.24–5.27 for ultraviolet D-radiation are similar to Figs. 5.20–5.23
for ultraviolet G-radiation.

The ultraviolet region 3150–3800 Å

In spite of the fact that this radiation represents roughly 90% of the
total ultraviolet energy, it has not been adequately studied because it was
considered to be of secondary importance. The total amount of this radiation
depends on $\gamma_\odot$ and increases by a factor of 5 from sunrise to noon (the
spectral composition of this radiation is shown in Fig. 5.11; reference should
also be made to Figs. 5.14, 5.15 and 5.16).

Fig. 5.28 illustrates the effect of the variation in $\gamma_\odot$ on the zenith D-
radiation, in relative units. The minimum at about 3900 Å is not, as mentioned above, a characteristic of the ultraviolet D-radiation, but of the I-
spectrum. The same is true for the maximum at about 4100 Å. The curves for
different $\gamma_\odot$ are in general similar, except for those corresponding to a very

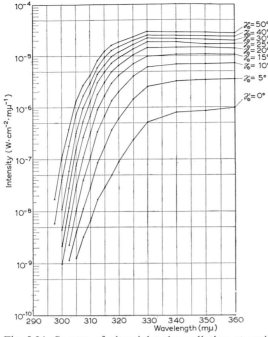

Fig. 5.24. Spectra of ultraviolet sky radiation at various solar altitudes $\gamma_\odot$ and for an amount $X = 0.250$ cm of atmospheric ozone [29].

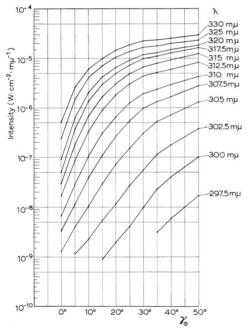

Fig. 5.25. The spectral intensity of ultraviolet sky radiation. Dependence on solar altitude for various wavelengths and for an amount $X = 0.250$ cm of atmospheric ozone [29].

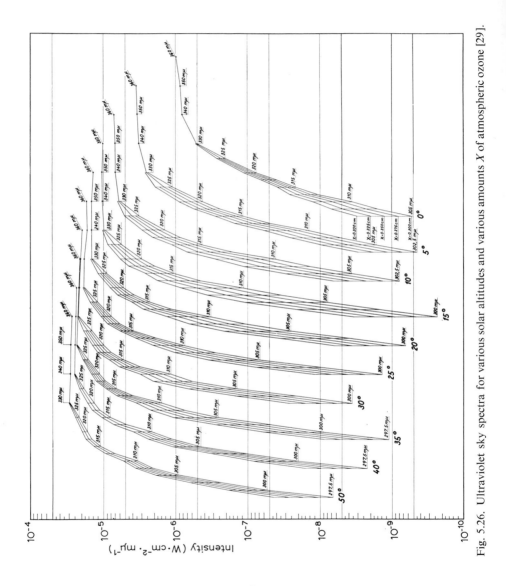

Fig. 5.26. Ultraviolet sky spectra for various solar altitudes and various amounts X of atmospheric ozone [29].

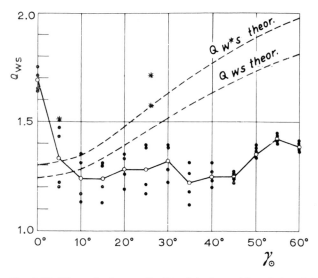

Fig. 5.27. The ratio $Q_{ws} = D_w/D_s$ of the intensities of ultraviolet sky radiation measured during winter and summer [29].
● Figures relating to $\lambda = 330$–360 mμ and to the mean intensities measured during the winter and summer periods.
○ Mean values, averaged over the interval 330 m$\mu \leqq \lambda \leqq$ 360 mμ.
* Figures Q_{w*s} relating to the first snowfall in October 1958.
- - - - - Theoretical values.

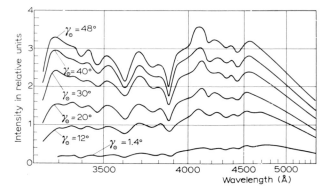

Fig. 5.28. Energy distribution in the ultraviolet zenith radiation [20].

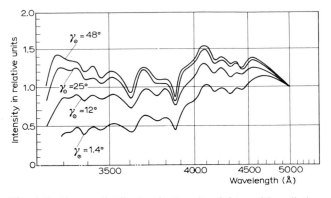

Fig. 5.29. Energy distribution in the ultraviolet zenith radiation, plotted relative to the energy at 5000 Å [20].

low sun. There is a relative maximum at 3275 Å and the energy falls off rapidly at shorter wavelengths.

Fig. 5.29 shows the relative distribution, the energy at 5000 Å being taken as unity.

It was mentioned above that the minimum wavelength λ_{min} occurs in the autumn because the thickness of the ozone layer reaches a minimum during that season. Fig. 5.30 shows that both the total amount of ultraviolet D-radiation and the magnitude of λ_{min} for this radiation are smaller in the autumn than in the summer. The autumnal curve in Fig. 5.30 corresponds to a transmissivity of 0.75 while the summer curve corresponds to a transmissivity of 0.72.

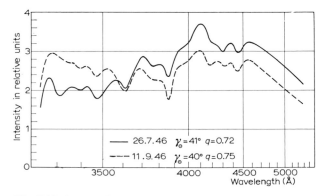

Fig. 5.30. Energy distribution in the ultraviolet zenith radiation on two dates [20].

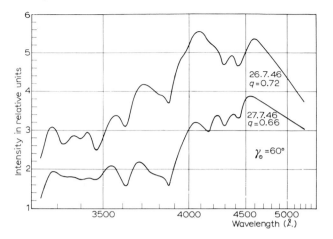

Fig. 5.31. Energy distribution in the ultraviolet zenith radiation on two different dates, but for the same $\gamma_\odot$ value [20].

Finally, Fig. 5.31 shows the energy distribution in the D-radiation coming from the zenith; the two curves are plotted for the same $\gamma_\odot$ value (60°) but corresponding to a different transmissivity.

The ultraviolet region 2800–3150 Å

This is the most important component of the ultraviolet radiation and will be discussed in some detail.

The entire energy of the ultraviolet radiation in this region, as a function of $\gamma_\odot$, is shown in relative units in Fig. 5.32.

The dependence of the ultraviolet *I-radiation* on the air mass is shown in Fig. 5.33. The combined effect of $\gamma_\odot$ and h on the I-radiation between 2800 and 3150 Å is illustrated in Table 5.9. The combined effect of h and of the season is shown in Table 5.10. A certain symmetry exists during the day (Table 5.9), except for the larger h values. The intensity is greater before noon than after noon, owing possibly to the increased turbidity in the afternoon. As may be seen in Table 5.10, winter values increase more rapidly with h than do the summer values; this is due to the purity of the atmosphere at high altitudes during the winter. Table 5.11 shows the combined effect of time and date on 2800–3150 Å I-radiation for a cloudless sky, in the lowland area. The values given are a.m. and p.m. averages on a horizontal plane, in relative units.

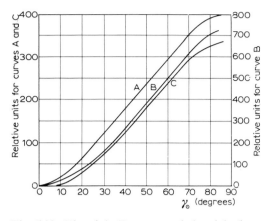

Fig. 5.32. Ultraviolet D-energy and ultraviolet I-energy as a function of $\gamma_\odot$ (2800–3150 Å, cloudless sky) [21].
(A) Ultraviolet D-radiation on a horizontal plane, (B) ultraviolet D- + I-radiation on a horizontal plane, (C) ultraviolet I-radiation on a plane normal to the beam.

As far as data are available, ultraviolet *D-radiation* in the range 2800–3150 Å may be treated in the same way as in the above cases. The dependence of this radiation on $\gamma_\odot$ at a horizontal plane is as shown by curve A in Fig. 5.32. The hourly (average) values of the intensity of the radiation on a horizontal plane are shown in Table 5.12 for a cloudless sky in relative units. There is only a very slight dependence on the altitude h for altitudes below 3300 metres.

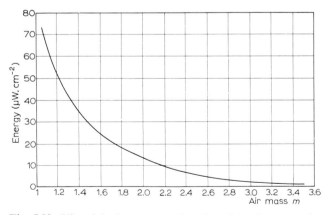

Fig. 5.33. Ultraviolet I-energy as a function of the air mass m ($\lambda \leqq 3132$ Å) [27].

TABLE 5.9. Ultraviolet I-radiation (2800–3150 Å): intensity in relative units for various values of $\gamma_\odot$ and h [22].

h (in m)	time	$\gamma_\odot$										
		15°	20°	25°	30°	35°	40°	45°	50°	55°	60°	65°
1180	a.m.	5.0	7.3	15.0	20.0	45.0	82	127	245	350	450	576
	p.m.	4.8	5.4	14.2	24.0	35.0	55	90	130	217	360	492
701	a.m.	—	5.0	13.0	15.0	33.0	35	89	170	255	330	466
	p.m.	—	—	11.0	17.0	28.0	40	75	150	175	250	360
470	a.m.	3.5	4.2	8.0	12.0	23.0	45	57	113	190	270	440
	p.m.	—	—	—	11.0	18.0	32	45	85	130	250	342
120	a.m.	1.8	3.6	5.0	7.0	15.0	28	40	85	143	242	280
	p.m.	1.8	3.8	4.6	6.0	12.0	15	28	54	85	128	242
0	a.m.	—	3.9	4.5	6.0	10.0	22	32	72	90	233	250
	p.m.	—	2.5	4.0	4.9	7.2	12	25	38	70	100	190

TABLE 5.10. Ultraviolet I-radiation (2800–3150 Å): intensity at different altitudes, relative to the intensity at $h = 200$ m [11, 21, 23, 24].

	h (in m)							
	200	500	1000	1500	2000	2500	3000	3500
Summer	100	125	145	170	182	190	195	200
Winter	100	150	220	280	330	390	440	480

TABLE 5.11. Ultraviolet I-radiation (2800–3150 Å): diurnal variation during the year [24].

Month	Hour							
	5	6	7	8	9	10	11	12
	19	18	17	16	15	14	13	
January	—	—	—	3	9	11	15	18
February	—	—	4	7	13	27	34	37
March	—	—	7	13	30	58	82	90
April	—	5	11	33	73	115	147	163
May	3	9	24	57	109	162	200	210
June	4	10	32	71	120	186	224	240
July	4	10	29	67	123	178	215	237
August	—	8	17	41	87	138	172	185
September	—	2	8	19	45	78	109	128
October	—	—	2	8	17	35	47	56
November	—	—	—	3	9	14	20	23
December	—	—	—	1	7	9	12	14

References pp. 194–195

TABLE 5.12. Ultraviolet D-radiation (2800–3150 Å): diurnal variation during the year [24].

Month	Hour							
	5 19	6 18	7 17	8 16	9 15	10 14	11 13	12
January	—	—	—	5	24	45	66	72
February	—	—	2	20	52	83	106	72
March	—	—	18	55	100	147	173	185
April	—	15	50	105	165	210	243	257
May	9	36	88	147	204	258	298	310
June	15	50	103	164	220	279	316	330
July	14	45	99	163	217	270	313	323
August	2	23	68	124	183	232	268	280
September	—	—	30	76	125	172	211	228
October	—	4	7	32	71	108	134	147
November	—	—	—	11	31	57	79	89
December	—	—	—	4	18	37	53	60

Fig. 5.32 also shows a plot of the combined solar and sky ultraviolet radiation (*G-radiation*) in the range 2800–3150 Å, on the horizontal plane, as a function of $\gamma_\odot$ for a cloudless sky (curve B). Daily amounts of this radiation throughout the year are shown in Table 5.13.

TABLE 5.13. Ultraviolet G-radiation (2800–3150 Å): diurnal variation during the year [24].

Month	Hour							
	5 19	6 18	7 17	8 16	9 15	10 14	11 13	12
January	—	—	—	8	33	56	81	90
February	—	—	6	27	65	110	140	155
March	—	—	25	68	130	205	255	275
April	—	20	61	138	238	325	390	420
May	12	45	112	204	313	420	498	520
June	19	60	135	235	350	465	540	570
July	18	55	128	230	340	448	528	560
August	2	31	85	165	270	370	440	462
September	—	6	38	95	170	250	320	356
October	—	—	9	40	88	143	181	203
November	—	—	—	14	40	71	99	112
December	—	—	—	5	25	46	65	74

TABLE 5.14. Ultraviolet insolation of horizontal and vertical planes of equal area [24].

Plane	March			July			October			December		
	I	D	G	I	D	G	I	D	G	I	D	G
Horizontal	25	50	75	133	132	265	43	89	132	2	14	16
Vertical	36	84	120	96	199	295	62	151	213	5	23	28

TABLE 5.15. Coefficient k (ratio of G-to I-radiation for $\lambda = 2800$–3150 Å) for various h values [21].

h (in metres)	0	500	1000	2000	3000
coefficient k	2.0	1.7	1.6	1.5	1.3

A vertical plane receives more radiation throughout the year than a horizontal plane (Table 5.14). The influence of surface orientation is usually described by a coefficient k equal to the ratio of G-radiation to I-radiation, both in the 2800–3150 Å range, the numerical values of which are given in Table 5.15 for different values of h; the maximum occurs at sea level.

In conclusion, it may be noted that for a cloudless sky, D-radiation in the 2800–3150 Å range is more intense than the I-radiation. In winter the latter increases by a factor of 6 during the day while the former increases by a factor of 14. In the summer, the corresponding values are 60 and 22.

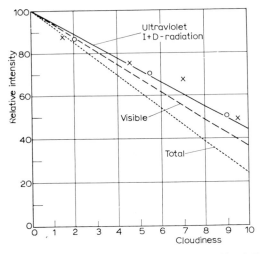

Fig. 5.34. The influence of clouds on combined I- + D-radiation in the range 2800–3150 Å compared with visible and total radiation [21].

As can be seen from Fig. 5.34, the ultraviolet radiation is more strongly affected by cloudiness than the radiation in other spectral regions. For further details the reader is referred to '*Medical Biometeorology*', edited by S. W. Tromp (Elsevier, Amsterdam, 1963), pp. 55–71.

6. THE VISIBLE REGION 3800–7800 Å

A distinction must be made between the total visible radiation and that part of radiation which is perceived by the eye, the latter being called 'light'. Two sensitivity curves are plotted in Fig. 5.35, curve 1 being for strong radiation, *e.g.* sunlight, and curve 2 for weak radiation, such as moonlight. It may be seen that curve 2 is displaced towards shorter wavelengths, *i.e.* the light appears bluer at low intensities. This is known as the Purkinje effect.

A detailed knowledge of the spectral composition of the visible radiation and of the possible changes due to various factors is very important, *e.g.* in photosynthesis. The spectral composition of visible I-radiation has already been discussed in connection with the overall I-spectrum (Figs. 5.4 and 5.5). Some additional data are summarized in Fig. 5.36 in which the energy is expressed in relative units (energy at 5600 Å being taken as unity). Curves 1, 2 and 3 were plotted from spectrophotometric data obtained at noon. Curves 4, 5 and 6 are computed on the basis of the spectrum for $m = 0$

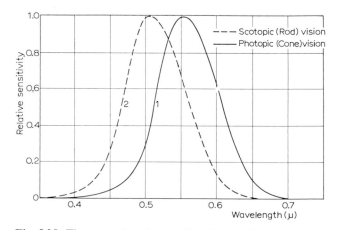

Fig. 5.35. The perception of strong (1) and weak (2) light.

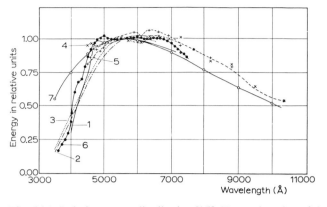

Fig. 5.36. Relative energy distribution [25]. For explanation of the curves, see text.

(outside the atmosphere) and on the transmissivity of the atmosphere under various conditions. Curves 4 and 5 correspond to $m = 2$ with the former taken on a clear dry day and the latter for a more humid atmosphere. In plotting curves 4 and 5, only scattering effects were taken into consideration. Curve 6 corresponds to $m = 1.59$ and an atmosphere of medium transmissivity, and finally curve 7 was computed for a black body at $5200°K$.

Fig. 5.37 shows plots of visible D-radiation (curve 1 is for a cloudless sky, curve 2 corresponds to the presence of clouds). For the purpose of comparison, two curves of visible I-radiation in the same spectral region are

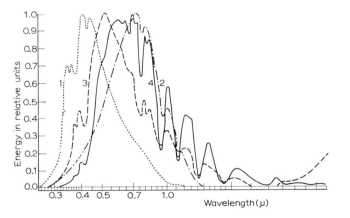

Fig. 5.37. I- and D-distributions for different values of the air mass m [26]. For explanation of the curves, see text.

References pp. 194–195

given (curve 3 is for $m=1.4$ and curve 4 for $m=4.0$). The curve for a cloudless sky is more similar to that for visible I-radiation for high sun, than to the D-curve for a cloudless sky.

Finally, let us consider the composition of the total I- + D-radiation in the visible range. There are two types of changes in the energy distribution, *i.e.* when the energy changes in the same proportion at all wavelengths, and when the energy changes selectively in various regions of the spectrum. The former case gives rise to a neutral effect, *i.e.* to no overall change in colour; the latter situation causes the colour balance to be altered. Both are of considerable importance in various applications, but details of these are outside the scope of this book.

7. THE SPECTRAL REGION 7800–14,000 Å

The influence of the air mass and altitude and various other factors on the spectrum in this region was discussed above in connection with the overall I-spectrum (Figs. 5.4 and 5.5). A characteristic feature of this spectral range is absorption by water vapour and carbon dioxide. The energy distribution is therefore very sensitive to changes in the atmosphere.

REFERENCES

1. N. N. KALITIN, *Actinometry* (in Russian), Leningrad (1938).
2. F. LINKE, *Handbuch der Geophysik*, Bornträger, Berlin, Vol. 8 (1942), pp. 246, 250.
3. F. E. FOWLE, *Smithsonian Inst. Misc. Collections*, 69, No. 3 (1918).
4. K. FEUSSNER AND P. DUBOIS, *Gerlands Beitr. Geophys.*, 27 (1930) 132–175.
5. B. M. DUGGAR (editor), *Biological Effects of Radiation*, McGraw-Hill, New York, Vol. 2 (1936).
6. F. LINKE, *Handbuch der Geophysik*, Bornträger, Berlin, Vol. 8 (1942), pp. 345, 353, 380.
7. P. HESS, *Gerlands Beitr. Geophys.*, 55 (1939) 211, 212, 216, 217.
8. R. HERMANN, *Thesis*, Giessen (1943), p. 391; *Strahlentherapie*, 76 (1947) 193.
9. R. HERMANN, *Thesis*, Giessen (1943), p. 392.
10. O. GLASER, *Medical Physics*, The Year Book Publ., Chicago, Ill.
11. A. E. H. MEYER AND E. O. SEITZ, *Ultraviolet Radiation*, Edwards, Ann Arbor, Mich. (1944).
12. R. STAIR, *J. Res. Natl. Bur. Std.*, 46 (1951) 355.
13. R. STAIR, *J. Res. Natl. Bur. Std.*, 49 (1952) 232.
14. R. STAIR, R. G. JOHNSTON AND TH. C. BAGG, *J. Res. Natl. Bur. Std.*, 53 (1954) 113.
15. E. A. POLYAKOVA, *Zh. Geofiz.*, 7, No. 2–3 (1937) 186–205.
16. W. E. FORSYTHE (editor), *Measurement of Radiant Energy*, McGraw-Hill, New York (1937).

17. L. GOLDBERG, *The Earth as a Planet* (edited by G. P. Kuiper), Univ. of Chicago Press, Chicago, Ill. (1954), p. 452.

18. N. ROBINSON, *U.V. in Haifa*, Internal Report, Solar Physics Laboratory, Israel Institute of Technology, Haifa.

19. F. LINKE, *Radioterap. Radiobiol. Fis. Med.*, 6 (1940) 37–41.

20. P. GÖTZ AND E. SCHÖNMANN, *Helv. Phys. Acta*, 21 (1948) 156.

21. K. BÜTTNER, *Physikalische Bioklimatologie* (Probleme der kosmischen Physik, Band 28), Akad. Verlagsges., Leipzig (1938).

22. V. K. ANISIMOV, *Meteorol. i Gidrol.*, No. 8 (1932).

23. I. ECKEL, *Meteorol. Z.*, 53 (1936) 90.

24. F. SAUBERER, *Wetter und Leben*, 7 (1955) 80–85.

25. M. V. SAVOSTYANOVA, *Izv. Akad. Nauk SSSR, Ser. Geol.*, No. 4 (1942).

26. J. D. YANISHEVSKY, *Tr. Gl. Geofiz. Observ.*, No. 26 (1951).

27. W. W. COBLENTZ AND R. STAIR, *J. Res. Natl. Bur. Std.*, 16 (1936) 315–317.

28. K. YA. KONDRATYEV, *Radiant Solar Energy* (in Russian), Leningrad (1954), p. 21.

29. P. BENER, *Contract AF61(052)–54*, Tech. Summary Rept. No. 1, Dec. 1960

Chapter 6

RADIATION BALANCE IN THE SYSTEM EARTH'S SURFACE-ATMOSPHERE

1. INTRODUCTION

The radiation balance in the system earth's surface–atmosphere is of special interest in dynamic meteorology. In this chapter we shall be concerned with the radiation balance B in this system and with its diurnal and seasonal variations. The geographical distribution of B will also be discussed. It should be noted that in general the radiation will penetrate to various depths below the earth's surface, but this will be neglected in order to simplify the discussion.

2. THE COMPONENTS OF THE RADIATION BALANCE

The radiation balance B consists of two main parts, namely, the radiation balance of the earth's surface B_E, which is made up of the gains and losses at the surface, and the corresponding radiation balance in the atmosphere B_A. Let us consider these two components in turn.

The radiation gains which constitute a part of B_E include the absorbed fractions of the incident direct solar radiation, the scattered sky radiation, and the thermal radiation emitted by the atmosphere (cf. Chapter 3). The radiation losses are made up of the radiation emitted by the surface and the reflected short-wave radiation. The overall radiation balance may be described by the following equation:

$$B_E = (Q_E + q_E)\,(1 - A_E) + \delta L_0\!\downarrow - L_0\!\uparrow \qquad (6.1)$$

where Q_E is the amount of incident direct solar radiation (I-radiation), q_E is the amount of incident scattered sky radiation (D-radiation), A_E is the albedo of the surface, which in general is a function of Q_E and q_E, $L_0\!\downarrow$ is the amount of terrestrial radiation emitted by the atmosphere and intercepted by the earth's surface, δ is the absorption coefficient of the surface for $L_0\downarrow$, and $L_0\uparrow$ is the amount of radiation emitted by the earth's surface in the upward direction.

It follows from the above definitions that $L_0\uparrow - \delta L_0\downarrow = L_0$ is the effective radiation of the surface, so that Eq. (6.1) may be rewritten in the form

$$B_E = (Q_E + q_E)(1 - A_E) - L_0 \tag{6.2}$$

The radiation balance of the atmosphere B_A consists of the proportion of I-radiation and D-radiation which is absorbed in the atmosphere on the way to the earth's surface and of the absorbed part of $L_0\uparrow$. This is the 'gain' component of B_A. The 'loss' component consists of the radiation emitted by the atmosphere in the direction of the earth's surface and into space ($L_0\downarrow$ and $L_\infty\uparrow$ respectively). The overall balance may therefore be described by

$$B_A = Q_A + q_A + \delta L_0 - (L_0\downarrow + L_\infty\uparrow) \tag{6.3}$$

where Q_A is the amount of I-radiation and q_A is the amount of D-radiation absorbed by the atmosphere.

The amount of thermal radiation absorbed by the atmosphere depends on the transmissivity P of the atmosphere for this particular type of radiation. The quantity $\delta L_0\uparrow$ may thus be written in the form $\delta L_0\uparrow = (1 - P)L_0\uparrow$. Moreover, from $L_0 = L_0\uparrow - \delta L_0\downarrow$, it follows that Eq. (6.3) may be rewritten in the form

$$B_A = Q_A + q_A + L_0 - L_\infty\uparrow \tag{6.4}$$

In order to obtain the total radiation balance in the system earth's surface–atmosphere, Eqs. (6.2) and (6.4) have to be added:

$$B = B_E + B_A = (Q_E + q_E)(1 - A_E) + Q_A + q_A - L_\infty \tag{6.5}$$

or

$$B = Q_0(1 - A_s) - L_\infty\uparrow \tag{6.6}$$

where A_s is the average albedo of the earth as a planet and Q_0 is the total flux of the solar radiation outside the atmosphere. The mean albedo A_s may be regarded as the sum of three components, *i.e.* $A_s = A_E + A_A + A_C$ where A_E is the albedo of the earth's surface (cf.Eq. (6.1)), A_A is the albedo associated with the back-scattering of the incident radiation by the atmosphere and A_C is the albedo of clouds (cf. Figs. 6.11 and 6.12).

3. THE EFFECT OF THE ALBEDO ON THE RADIATION BALANCE

The role of the albedo is clear from the above equations for B_E, B_A and B_s. Table 6.1 gives some typical average values of the albedo for various kinds of soil. Experimental methods for the determination of the albedo are considered in Chapter 7.

TABLE 6.1. Values of A_E for various kinds of soil [1].

Kind of soil	A_E (%)
Black soil, dry	14
Black soil, moist	8
Grey soil, dry	25–30
Grey soil, moist	10–12
Blue loam, dry	23
Blue loam, moist	16
Fallow, dry	8–12
Fallow, moist	5–7
Ploughed field, moist	14
Desert, loamy surface	29–31
Sand, yellow	35
Sand, white	34–40
Sand, river	43
Sand, bright, fine	37

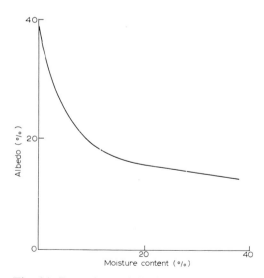

Fig. 6.1. Dependence of albedo on the moisture content of sand [1].

TABLE 6.2. Influence of surface roughness on A_E [1].

Kind (roughness) of the surface	A_E (%)
Flat, even	30–31
Covered with dust	28
Covered with pellicles, after moistening	27
Covered with small pebbles	25
Covered with bigger pebbles	20
Newly ploughed field	17

Since the numerical data given in Table 6.1 represent experimental averages, it is clear that individual measurements may deviate quite considerably from these data. The term 'soil' must be defined in terms of the amounts of moisture, the grain size (roughness of surface) and the colour of the surface. The albedo of all kinds of soil decreases with increasing amount of moisture because the albedo of water is lower than that of soil. Fig. 6.1 shows an example of this in the case of sand. The effect of the surface texture is indicated by the data in Table 6.2.

The albedo of a vegetation cover depends on the type of vegetation and on its age. Some examples of this are given in Table 6.3. Fig. 6.2 shows the change in the albedo for a number of crops.

The albedo of snow depends on the amount of impurity it contains, on its surface roughness and on the angle of incidence. Some examples of this dependence are shown in Table 6.4. Table 6.5 shows the albedo of ice as a function of the state of its surface.

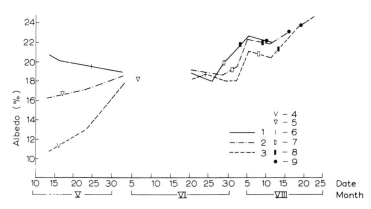

Fig. 6.2. Changes of the albedo of rye and wheat with the vegetation period [1].
(1) Winter rye, (2) winter wheat, (3) spring wheat, (4) beginning of earing, (5) full tube, (6) full earing, (7) milky ripeners, (8) wax ripeners, (9) full ripeners.

The values of the albedo given above are averages over various kinds of surface cover and also averages with respect to time. In reality the albedo exhibits a diurnal and an annual variation, which was to be expected in view of the variations in the elevation of the sun. The albedo is found to increase with decreasing elevation of the sun. An example of the diurnal variation in the albedo is shown in Fig. 6.3.

TABLE 6.3. Influence of vegetation cover on A_E [1].

Vegetation	A_E (%)
Spring wheat	10–25
Winter wheat	16–23
Winter rye	18–23
High grass, densely grown	18–20
Green grass	26
Grass dried in the sun	19
Tops of oak	18
Tops of pine	14
Tops of fir	10
Cotton	20–22
Lucerne (beginning of blossoming)	23–32
Rice field	12
Lettuce	22
Beets	18
Potatoes	19
Heather	10

TABLE 6.4. Values of A_E for various types of snow cover [1].

State of surface	Sun elevation (degrees)	A_E (%)
Dense, dry and clean snow	30.3	86
	29.7	88
	25.1	95
Clean, moist, fine-granular snow	33.3	64
	34.5	63
	35.5	63
Clean, moist, granular snow	33.7	61
	32.0	62
Porous, very moist, greyish snow	35.3	47
	36.3	46
	37.3	45
Snow cover of sea ice, very porous, grey, soaked by water	32.8	43
	31.7	43
Very porous, light brown snow, soaked by water	29.7	31
Very porous, dirty snow, soaked by water, sea ice seen through cover	37.3	29

TABLE 6.5. Dependence of A_E of ice on the state of the surface [1].

State of surface	A_E (%)
Sea ice, slightly porous, milky bluish	36
Melting ice, very porous	41
Ice with light snow cover	31
Ice sheet, covered by a water layer of 15–20 cm	26
White porous ice layer on sea	12

Isopleths of some common forms of surface cover are given in Fig. 6.4. It is evident that there is an increase in the albedo with decreasing elevation of the sun and a minimum is reached near noon.

As has already been mentioned, the albedo exhibits an annual variation which is intimately connected with changes in the surface cover. Fig. 6.5 shows an example of this variation. It is clear from this figure that the values of the albedo at a particular point may be different for different years, depending on the amount of snow, rain, etc.

So far we have discussed the albedo of solid surfaces with various types of cover. Let us now consider the albedo of water surfaces. It has already been mentioned in connection with Eq. (6.1) that the values of the albedo are different for direct solar radiation and for scattered sky radiation. This difference is particularly noticeable in the case of water surfaces and we shall, therefore, discuss these two cases separately.

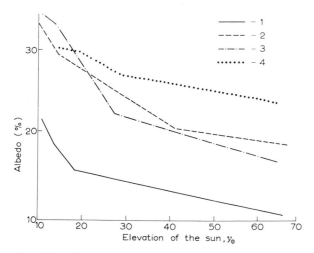

Fig. 6.3. Daily course of A for various coverages [1].
(1) Stony soil, (2) partly dry grass, (3) loam, (4) dry greyish-green soil.

Let us consider the albedo for I-radiation (A_I) first. For a smooth water surface, the Fresnel formula may be employed, and the corresponding values of A_I have been computed for various values of the angle of incidence. The annual values of A_I are very dependent on the roughness of the water surface. For very smooth sea surfaces the experimental values of A_I are in good agreement with the theoretical calculations. This can be seen from Table 6.6.

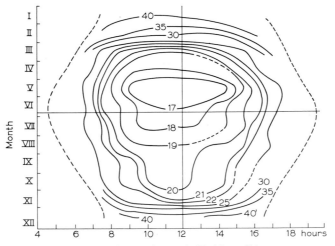

Fig. 6.4. A-isopleths of natural cover in Tashkent [1].

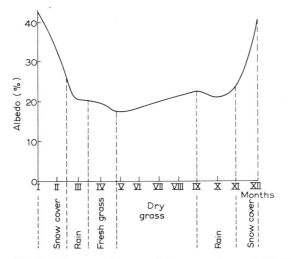

Fig. 6.5. The annual course of A for natural cover in Tashkent [1].

TABLE 6.6. Computed and observed values of A_I for the surface of the sea [1].

Sun elevation (degrees)	A_I (%)		
	Computed (Fresnel)	Observed (Kuzmin)	Observed (Ångström)
90	2	—	—
70	2	—	—
50	2.5	—	—
30	6	6	8
25	9	8	10
20	14	12	15
15	21	18	28
10	35	32	49
8	43	40	56
6	53	48	70
4	65	60	—

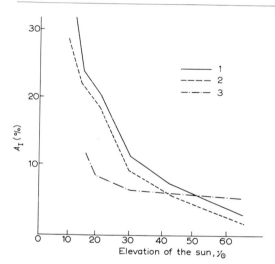

Fig. 6.6. Daily course of A_I of a sea [1].
(1) Muddy water, (2) slightly muddy water, (3) clear water.

There are at least three reasons for the discrepancy between the computed and measured values of A_I. Thus, a portion of the reflected D-radiation is mixed with the reflected I-radiation, the scattered reflected radiation from the water surface may be included in the measurement, and the roughness of the surface may contribute significantly to the effect. A further factor is the transmissivity of the water. Fig. 6.6 shows the diurnal variation in A_I for three values of the transmissivity.

A further factor which affects the magnitude of A_I is the refractive index of water n. It is found that there is a rapid increase in A_I with increasing n. This is of particular importance because as A_I increases, less energy is absorbed for evaporation. For example, in the case of the Dead Sea, the refractive index is $n = 1.391$ (at 27°C).

TABLE 6.7. Influence of roughness of the surface of the sea on A_I [1].

Zenith distance $z_\odot$	A_I of sea water (%)	
	Rough surface	Smooth surface
0	13.1	2.1
30	3.8	2.2
60	2.4	6.2

Table 6.7 shows a comparison of theoretical values of A_I for smooth and rough water surfaces, at various zenith distances. In the computation of the values for rough water surfaces the assumption was made that the form of the waves was such that they have a circular cross-section.

The above discussion of the albedo of water surfaces will be concluded with a short survey of the albedo of such surfaces for the scattered sky radiation A_D. The measurements and theoretical calculations of A_D are rather complicated. One of the reasons for this is the anisotropy in the luminance of a cloudless sky. The effect is particularly significant when the sky is partly clouded. Another reason is the low intensity of D-radiation as compared with I-radiation. In view of this, considerable differences between the computed and measured values of A_D are to be expected. Thus, experimental determinations have led to average values between 4.8 and 8.1%, whilst the computed results vary from 6.5 to 5.6% [1,2].

Table 6.8 shows the computed values of A_D for midday in the Crimea. These results are based on D-radiation measurements for an anisotropic sky

TABLE 6.8. Values of A_D for the surface of the sea as computed from observational data [1].

Date of observation	State of sky	A_D (%)
June 28, 1951	clear	8.5
June 30, 1951	clear	8.0
June 29, 1951		9.9
July 4, 1951		11.1

luminance. These values of A_D are larger than those for an isotropic sky. The difference between the two sets of results reaches a maximum in the case of a cloudy horizon.

4. THE EFFECT OF THE SPECTRAL COMPOSITION OF THE INCIDENT RADIATION ON THE ALBEDO

In general, the albedo A depends on the spectral composition of the incident radiation. The diurnal variation in A, which is indicated in Figs. 6.3 and 6.4, is only partly due to the dependence of A on the angle of incidence and the reflectivity of the surface. A further factor which is responsible for this variation is the fact that the incident radiation contains a higher proportion of short-wave radiation as the elevation of the sun increases (cf.

TABLE 6.9. A_λ of some natural surfaces [1].

Kind of surface	A_λ (%)					
	U.V. 3800 Å	Blue 4500 Å	Green 5500 Å	Red 6120 Å	Dark red 6600 Å	Cloudiness
Dug vineyard	4.9	16	23	27	16	7, Ci
Grass	1.4	6	15	15	24	7, Ci
Low, dry grass	3.3	7.9	18	18	20	7, Ci
Shore, dry	11	34	36	37	20	3, Fr. cu
Shore, wet	5	17	20	23	12	3, Fr.cu
Fine sand, dry	11	27	29	30	18	3, Fr.cu
Fine sand, wet	5.2	12	14	16	9.2	3, Fr.cu
Coverage by old pine needles in a pine forest	15	34	34	32	18	9, St
Sifted sea shore sand	6.4	19	20	20	15	10, Cs

TABLE 6.10. A_λ of various surfaces on clear days at noon [1].

General characteristics of the surface	State of the soil	A_λ (%)				
		Integral	Blue 4690 Å	Green 5320 Å	Red 6420 Å	Infrared 10400 Å
A section of a cotton field with vegetation	After watering	16	4	8	9	26
The same, without vegetation	After watering	15	8	10	12	22
The same, without vegetation, soil dug	Dry	21	11	18	20	28

Chapter 3). The effect of the wavelength dependence on A shows a trend towards an increase in A with decreasing elevation of the sun.

Tables 6.9 and 6.10 show some results of spectral albedo measurements. In practice, the wavelength range is defined by means of special filters. The integral albedo A is then obtained by integrating the spectral albedo A_λ, obtained in this way, over the entire spectrum.

5. DIURNAL VARIATION OF THE RADIATION BALANCE AND ITS COMPONENTS

Fig. 6.7 shows an example of the diurnal variation in B and of its components.

It is found that the maximum positive value of B occurs at about noon and the maximum negative value is found at night. The transition from positive to negative values of B does not correspond exactly with sunrise and sunset. In the morning, the transition from negative to positive values of B is delayed by about an hour, whilst the time of the transition from positive to negative values of B precedes sunset by about 1.5 hours. The first effect is due to the fact that a finite time interval is necessary for the incoming radiation to compensate for losses caused by the effective radiation emitted by

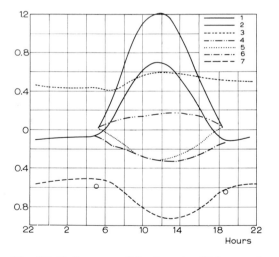

Fig. 6.7. Daily course of the elements of B (mean values for clear days) [3].
(1) B, (2) I, (3) $\delta L_0 \downarrow$, (4) D, (5) reflected short-wave radiation, (6) L_0, (7) $L_0 \uparrow$.

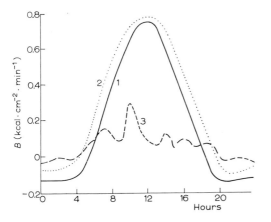

Fig. 6.8. Daily course of B [3].
(1) In Tashkent, mean for June, (2) in Koltushy (Leningrad zone), for a clear day in July, (3) in Koltushy, for a cloudy day in July.

the surface. The second effect is the consequence of the fact that the effective outward radiation exceeds the incoming radiation before sunset.

The daytime values of B are subject to greater variations than the night values. The diurnal distribution curve is asymmetric: it is lower in the afternoon than in the morning. This is due to the increase in the effective surface radiation in the afternoon. The diurnal variation shown in Fig. 6.7 refers to a cloudless day. The presence of clouds gives rise to a reduction in B. In fact, B is a rapidly varying function of the cloudiness, as indicated by Fig. 6.8.

The diurnal variation of B is also affected by other elements, such as

TABLE 6.11. Daily course of B on watered field and a semi-desert [3].
B in cal $\cdot$ cm^{-2} $\cdot$ min^{-1}.

Hour	Cotton field, watered	Semi-desert
0–1	−0.07	−0.08
4–5	−0.06	−0.06
6–7	0.18	0.13
8–9	0.62	0.47
10–11	0.89	0.66
12–13	0.96	0.66
14–15	0.87	0.56
16–17	0.46	0.30
18–19	0.03	−0.05
20–21	−0.07	−0.10

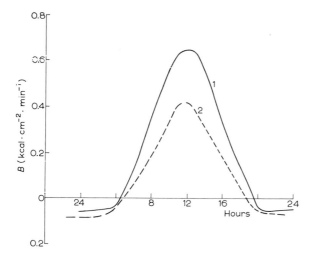

Fig. 6.9. Daily course of B [3].
(1) On bare soil, (2) soil covered by vegetation.

vegetation. This is shown in Table 6.11 and Fig. 6.9, where curve 1 refers to a bare soil and curve 2 to a soil covered by vegetation.

It is evident from these data that at night there is almost no difference between the values of B for the two types of surface. On the other hand in daytime the surface of a watered cotton field has a much higher B than a bare surface, which may be partly explained by the higher effective radiation emitted by the bare surface. The effect of watering on B may be described by

$$\Delta B = (Q + q)(A - A') + \beta(t_0 - t'_0) \tag{6.7}$$

where ΔB is the change in B, A and A' are the values of the albedo before and after watering, t_0 and t'_0 are the surface temperatures before and after watering, and $\beta = 0.008$ [3].

A further example of B which we shall consider is that of snow. The value of B for snow is almost constant over 24 hours, being in fact very nearly zero. This is explained by the high values of A and L_0 for snow.

6. SEASONAL VARIATION OF THE RADIATION BALANCE AND ITS GEOGRAPHICAL DISTRIBUTION

The seasonal variation of B is illustrated by Tables 6.12 and 6.13 which refer to a region near Leningrad and to the White Sea and the Atlantic Ocean respectively (cf. also Fig. 6.5).

The geographical distribution of the yearly sums of B for continents

TABLE 6.12. Annual course of monthly sums of B and of its components (near Leningrad) [3].

Values in kcal $\cdot$ cm^{-2} $\cdot$ month^{-1}.

Month	Total radiation	Absorbed radiation	L_0	B
January	0.8	0.2	2.1	-1.9
February	2.1	0.6	2.1	-1.5
March	5.4	2.7	3.1	-0.4
April	9.5	8.5	4.2	4.3
May	13.6	10.9	4.8	6.1
June	14.6	11.7	4.7	7.0
July	14.2	11.3	4.7	6.6
August	10.0	8.0	3.7	4.3
September	6.4	5.1	3.3	1.8
October	3.0	2.7	2.8	-0.1
November	1.0	1.0	2.0	-1.0
December	0.5	0.2	1.8	-1.6
Year	81.2	62.9	39.3	23.6

TABLE 6.13. Annual course of monthly sums of B for the White Sea and the Atlantic Ocean [3].

B in kcal $\cdot$ cm^{-2} $\cdot$ month^{-1}.

Month	Place of observation	
	White Sea	Atlantic Ocean (40°N, 40°W)
January	-3.2	-0.5
February	-2.2	1.6
March	-0.8	4.6
April	1.6	6.6
May	5.6	9.2
June	6.9	9.7
July	5.7	10.2
August	3.2	9.1
September	1.2	6.9
October	-1.3	3.0
November	-2.7	0.8
December	-3.3	-0.8
Year	10.7	60.4

TABLE 6.14. Mean latitudinal distribution of annual sums of B on continents, oceans and the total earth's surface [3].
B in kcal · cm^{-2} · year^{-1}.

Geographical latitude, $\varphi°$	Oceans	Continents	Total earth's surface
60–50 N	34	23	28
50–40 N	54	38	46
40–30 N	78	56	69
30–20 N	100	64	86
20–10 N	110	74	101
10–0 N	107	79	101
0–10 S	107	75	99
10–20 S	107	69	99
20–30 S	94	62	87
30–40 S	73	55	71
40–50 S	53	39	53
50–60 S	31	26	31
Total earth's surface	77	46	68

and oceans and the total earth's surface is given in Table 6.14 and on the chart of Fig. 6.10.

It is interesting to note that the yearly sums are always positive. The variation curves show sharp changes at the boundaries between continents and oceans. On the oceans, B-lines are almost coincident with the lines of equal geographical latitude. There is a reduction in B as the North and South Pole are approached.

7. THE RADIATION BALANCE OF THE ATMOSPHERE

It is evident from our discussion of absorption in Chapter 3 that $Q_A + q_A$ in Eq. (6.4) is small compared with L_0 and $L_\infty\uparrow$. It is also clear that $L_\infty\uparrow$ is greater than L_0, which means that $B_A < 0$. We shall assume that the atmosphere is a uniform participant in the radiation balance, *i.e.* that B_A is independent of height. Table 6.15, which is based on this assumption, shows that the annual variation of B_A is a function of geographical latitude.

It is clear from these data that at northern latitudes B_A decreases between the equator and the latitude of 25°, increases until the latitude of 60° is reached, and decreases again with increasing latitude thereafter. The

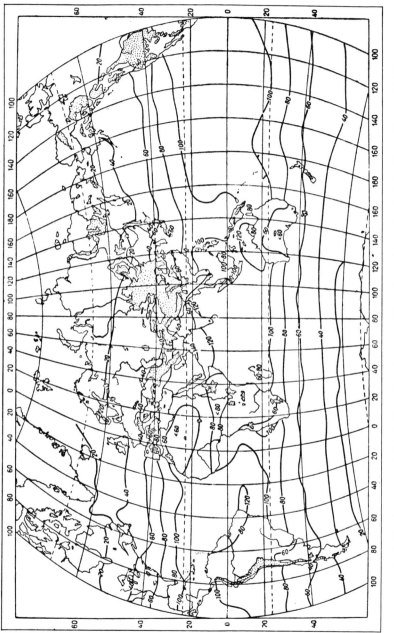

Fig. 6.10. Geographical distribution of yearly sums of B on the earth's surface (B in kcal · cm^{-2} · year^{-1}) [3].

TABLE 6.15. Latitudinal distribution of annual sums of B_A in the northern hemisphere [3].
B_A in kcal · cm^{-2} · year^{-1}.

Geographical latitude, $\varphi°$	B_A
0–10 N	−56
10–20 N	−54
20–30 N	−50
30–40 N	−59
40–50 N	−69
50–60 N	−76
60–90 N	−73
0–90 N	−60

mean annual value of B_A for the entire northern hemisphere is 60 kcal · cm^{-2} · year^{-1}. The values of the components in this case are $L_0 = 50$, $L_\infty{\uparrow} = 145$, and $Q_A + q_A = 35$ (all in kcal · cm^{-2} · year^{-1}).

8. THE ALBEDO OF CLOUDS

The albedo of clouds, which will clearly affect the radiation balance B, must be measured from above the clouds. The albedo is found to be very

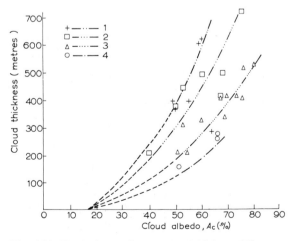

Fig. 6.11. Dependence of A_C on cloud thickness [1].
(1) Sc-10, broken, (2) Sc-Cu, (3) Sc-10, (4) Ac-10.

dependent on cloud thickness and this dependence is different for the various types of cloud. Fig. 6.11 shows some results of experimental determinations.

Each point in this figure is an average of forty measurements on a totally overcast sky. The curves in Fig. 6.11 do not pass through the origin because at zero cloud thickness the albedo is that of the earth's surface. Fig. 6.12 shows experimental determinations at Moscow (curve 1) and in Cali-

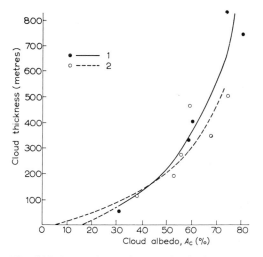

Fig. 6.12. Dependence of A_C on cloud thickness.
(1) In Moscow [1], (2) in California [4].

fornia (curve 2). In spite of the small number of points and their scattering (both in Moscow and California), they represent the general trend of a non-linear increase in the albedo with the thickness of the clouds.

The mean value of the albedo for clouds of various types and thickness is 0.50–0.55.

9. THE SPECTRAL ALBEDO OF THE EARTH AND ATMOSPHERE

As has already been pointed out, the albedo is a function of wavelength. In this section we shall divide the incident radiation into three spectral regions, namely $\lambda < 4000$ Å (ultraviolet), 4000 Å $\leqslant \lambda \leqslant 7400$ Å (visible) and $\lambda > 7400$ Å (infrared). We shall use the mean radiation incident on the earth, S_0, as a basis for further computations. Measurements outside the atmosphere (cf. Chapter 1, Table 1.2) have shown that the total radiation S_0 may be resolved into ultraviolet (9%), visible (45%) and infrared (46%). For the subozonal radiation (cf. Chapter 3) the above contributions to S_0 in the three spectral ranges are respectively 7, 45 and 46%, since about 2% of the ultraviolet is absorbed by the ozone layer.

According to a report from the Smithsonian Institution [5], $A_E = 2.3\%$ of S_0.

The albedo of the atmosphere A_A may be resolved into three components:

$$A_A = A_a + A_w + A_d \qquad (6.8)$$

where A_a is the contribution due to scattering by air, A_w is the component associated with scattering by water vapour and water droplets, and A_d is due to scattering by larger particles, for example, dust (cf. Chapter 3). Reported values for these quantities [5,6] are $A_a = 15\%$, $A_w = 10\%$ and $A_d = 5\%$ of S_0.

The relative proportions of the three contributions to A_A as given by Eq. (6.8) depend on the type of scattering, $i.e.$ on the ratio between wavelength of the scattered radiation and the size of the scattering particles (cf. Chapter 3). One approach is to assume that scattering by air or by water vapour and water droplets is symmetric, so that equal amounts are scattered in the upward and downward directions. Scattering by dust particles is asymmetric, the upward and downward scattering components being $\frac{1}{4}$ and $\frac{3}{4}$ of the total, respectively [6]. This means that the contribution of A_d to A_A is 1% of S_0, so that the total amounts to $8 + 5 + 1 = 14\%$ of S_0.

The albedo A_A must also be resolved into two parts: the fraction of A_A that is due to the cloudless atmosphere and the fraction due to the clouds. If we assume an average cloudiness of 0.54, then the mean value of the albedo of a cloudless atmosphere is $0.46 \times 14 = 6.4\%$ of S_0. The radiation scattered by air above the clouds in the upward direction must be added to this figure. This radiation amounts to $0.54 \times 5 = 2.7\%$ and hence the total albedo of the atmosphere is $6.4 + 2.7 = 9.1\%$ of S_0.

Finally, distribution of energy between the abovementioned three spectral regions must be considered. This distribution is approximately 10% (ultraviolet), 20% (visible) and 5% (infrared), and the spectral albedo of the atmosphere for these three regions is 2.6, 5.2 and 1.3% of S_0, respectively.

It follows from the foregoing discussion that the sum of the albedos of the earth's surface and of the atmosphere in the visible region of the spectrum is $5.2 + 1.1 = 6.3\%$ of S_0.

It follows that the albedo of the earth's surface and of the atmosphere in the visible part of the spectrum is only $6.3/0.45 = 14\%$, since the visible region contains 45% of the total radiant energy. Measurements of the luminance associated with the lunar ashen light have led to a figure of 39% for the visible part of the spectrum [7].

10. THE SPECTRAL ALBEDO OF THE CLOUDS

It follows from the foregoing discussion that the albedo of the clouds in the visible range is $39 - 14 = 25\%$ of the value of S_0 in this region, and this is only $25 \times 0.45 = 11.25\%$ of the total value of S_0. Theoretical calculations indicate that the ultraviolet component of the albedo of the clouds is 1.8% while the infrared component amounts to 10.2% of S_0. Thus, we find that $A_C = 1.8 + 11.25 + 10.2 = 23.25\%$. The total albedo of the earth including clouds is therefore $2.3 + 9.1 + 23.25 = 34.65\%$. The components of the albedo of the earth as a planet are indicated in Table 6.16.

TABLE 6.16. Spectral composition of A_s (in %) [7].

	Ultraviolet	Visible (v) range		Infrared	Total
		$S_{0,v}$	S_0		
Earth's surface	0.1	2.4	1.1	1.1	2.3
Clouds	1.8	25.1	11.3	10.2	23.3
Atmosphere	2.6	11.5	5.2	1.3	9.4
Total value of A	4.5	39.0	17.6	12.6	34.7
Spectral	50.0		39.0	27.0	35.0

The actual value of A_C is greater than 23.25% because the total radiation reaching cloud tops is $0.9\,S_0$ rather than S_0. The true albedo of the clouds is therefore $23.25/(0.9 \times 0.54) = 48\%$.

All the above data are based on the assumption that the scattering by air, water vapour and water droplets is symmetric with respect to the horizontal plane, while the scattering by dust in the upward and downward directions is in the ratio of $1:4$. Other assumptions concerning the scattering by water vapour, water droplets and dust can also be made. For example, a possible assumption is that scattering by water vapour, water droplets and by dust is negligible. If this assumption is in fact accepted, then the result for the cloudless part of the atmosphere becomes $8 \times 0.46 = 3.7\%$ of S_0. A figure

TABLE 6.17. Spectral composition of A_s (in %), assuming pure molecular scattering [7].

	Ultraviolet	Visible (v) range		Infrared	Total
		$S_{0,v}$	S_0		
Earth's surface	0.1	2.4	1.1	1.1	2.3
Clouds	2.0	28.6	12.9	11.6	26.5
Atmosphere	2.4	8.0	3.6	0.4	6.4
Total value of A	4.5	39.0	17.6	13.1	35.2
Spectral	50.0		39.0	29.0	35.0

of 2.7% of S_0 has to be added to the latter in order to correct for the radiation intercepted by the atmosphere above the clouds. The final result is therefore $3.7 + 2.7 = 6.4$%. Since the latter figure is due to pure molecular scattering, the spectral distribution of this radiation may be computed from the Rayleigh scattering formula (cf. Chapter 3). The results of this calculation are: ultraviolet 2.4%, visible 3.6%, and infrared 0.4%. Table 6.17 gives some numerical results of this calculation. This may be compared with the case considered above. A somewhat higher result was found for A_C so that we shall use $A_C = 50$–55%.

11. THE SEASONAL VARIATION AND THE GEOGRAPHICAL DISTRIBUTION OF THE MEAN ALBEDO OF THE EARTH AND ITS COMPONENTS

Table 6.18 shows an example of the seasonal variation of A_s in the northern hemisphere, for cloudless sky.

The increase in the share of A_E towards the winter is due to the increase in the snow cover and to a lower sun. The slight variation in the radiation scattered back by the atmosphere is due to the compensation of the increase in the water vapour content at low latitudes in summertime, which is associated with a higher sun. Fig. 6.13 shows an example of the seasonal distribution of A_s.

TABLE 6.18. Seasonal distribution of A_s and its components (in %) (mean values for a cloudless sky in the northern hemisphere) [8].

Components	Winter	Spring	Summer	Autumn
Reflected from the earth's surface, A_E	9.0	7.4	5.0	6.0
Scattered by the atmosphere	8.3	7.6	8.5	7.4
A_s	17.3	15.0	13.5	13.4

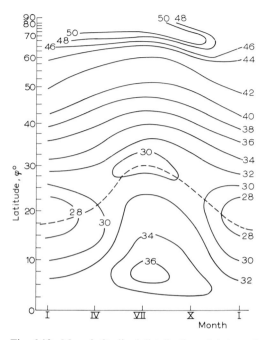

Fig. 6.13. Mean latitudinal distribution of A_s in various seasons [8].

It is clear from Fig. 6.13 that the minimum of A_s occurs in the sub-tropical region which is the region of minimum cloudiness. In the summer, the minimum shifts in the northward direction, whilst an opposite effect is found to occur in the winter. A_s is found to increase in the northward direction throughout the entire year up to northern latitudes of 60–70°, because of increasing cloudiness, increasing back-scattered radiation and reflection from the earth's surface with decreasing elevation of the sun. The geographical distribution of the mean annual values of A_s is shown in Fig. 6.14. This map was constructed on the basis of the formula

$$A_s = (A_1 + A_2)(1-n) + (A_3 + A_2')n \qquad (6.9)$$

where A_1 is the albedo of the earth's surface, A_2 is the albedo of the atmospheric layer between the earth's surface and the lower boundary of clouds, A_3 is the albedo of clouds, A_2' is the albedo of the layer of the atmosphere above the clouds, and n is the degree of cloudiness.

TABLE 6.19. Seasonal distribution of mean values of B in the northern hemisphere and of its components under conditions of mean cloudiness [8].
B in kcal · cm^{-2} · min^{-1}.

Components of A_s	Winter	Spring	Summer	Autumn	Year
I. Incoming short-wave radiation					
1. Insolation on the top of the atmosphere	0.348	0.580	0.645	0.424	0.500
2. Absorption of radiation in the atmosphere					
a. by ozone	0.011	0.016	0.019	0.010	0.014
b. by water vapour and dust	0.044	0.067	0.092	0.057	0.065
c. by clouds	0.005	0.010	0.010	0.007	0.008
Total absorption	0.060	0.093	0.122	0.074	0.087
3. Reflection and scattering of radiation into space					
a. by the atmosphere	0.023	0.037	0.048	0.028	0.034
b. by clouds	0.078	0.141	0.162	0.103	0.121
c. by the earth's surface	0.015	0.029	0.024	0.018	0.021
Total reflection	0.116	0.207	0.234	0.149	0.176
4. Absorption of radiation by the earth's surface					
a. direct sun radiation (I)	0.085	0.142	0.129	0.091	0.112
b. after passage through the clouds	0.045	0.091	0.090	0.064	0.072
c. scattered radiation (D)	0.043	0.050	0.070	0.048	0.053
Total absorption by the earth's surface	0.173	0.283	0.289	0.203	0.237

12. SEASONAL VARIATION IN THE RADIATION BALANCE OF THE SYSTEM EARTH'S SURFACE–ATMOSPHERE AND ITS COMPONENTS

The seasonal variation of the mean value of this balance and of its components in the northern hemisphere is given in Table 6.19 for average cloudiness.

An example of the geographical distribution of the mean annual values of the balance and its components is given in Table 6.20. It is clear from this table that the radiation balance is a function of latitude, and chan-

TABLE 6.19. (continued)

Components of A_s	Winter	Spring	Summer	Autumn	Year
II. Long-wave radiation					
1. Effective radiation of the earth's surface					
a. thermal radiation of the earth's surface	0.530	0.564	0.614	0.581	0.572
b. thermal radiation of the atmosphere	0.439	0.473	0.523	0.494	0.482
c. effective radiation	0.091	0.091	0.091	0.087	0.090
2. Thermal radiation of the troposphere					
a. thermal radiation absorbed by the troposphere	0.501	0.535	0.588	0.555	0.545
b. own thermal radiation of the troposphere	0.716	0.749	0.817	0.778	0.765
c. balance of the long-wave radiation of the troposphere	0.215	0.214	0.229	0.223	0.220
3. Thermal radiation into space					
a. of the earth's surface (inside the windows of transparency)	0.029	0.029	0.026	0.026	0.027
b. of the troposphere	0.277	0.276	0.294	0.283	0.283
c. of the stratosphere	0.011	0.016	0.019	0.010	0.014
Total outgoing radiation	0.317	0.321	0.339	0.319	0.324

TABLE 6.20. Latitudinal distribution of annual mean values of B and its components in the northern hemisphere [8].

Values in $kcal \cdot cm^{-2} \cdot min^{-1}$.

Geographical latitude, $\varphi°$	Absorbed solar radiation	Outgoing radiation	B
0–10 N	0.403	0.347	0.056
10–20 N	0.409	0.354	0.055
20–30 N	0.387	0.353	0.034
30–40 N	0.341	0.327	0.014
40–50 N	0.276	0.306	−0.030
50–60 N	0.224	0.287	−0.063
60–70 N	0.169	0.270	−0.101
70–80 N	0.122	0.253	−0.131
80–90 N	0.160	0.245	−0.139
0–90 N	0.324	0.324	0.000

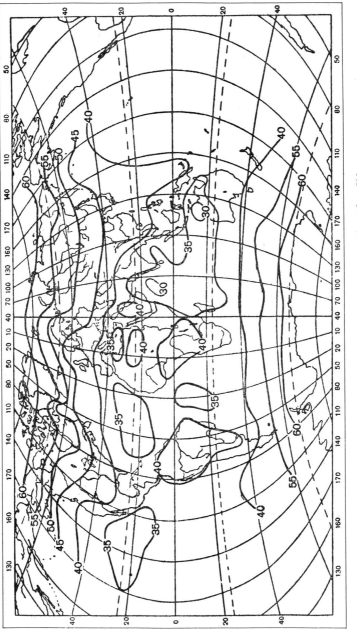

Fig. 6.14.   Mean annual values of A_s on the earth's surface [8].

ges from positive to negative values between 40° and 50°N. The geographical distributions in the summer and winter periods are appreciably different from that for the annual mean values. Thus, for example, in the winter the change of sign occurs near 20°N, while in the summer it is observed to occur at 80°N. Short-period changes in the balance are found to be even larger than those indicated above.

REFERENCES

1. K. YA. KONDRATYEV, *Radiant Solar Energy* (in Russian), Leningrad (1954), pp. 441–483.
2. H. GRIESSEIER, *Z. Meteorol.*, 6 (1952) 2.
3. K. YA. KONDRATYEV, *Radiative Heat Transfer in the Atmosphere* (in Russian), Leningrad (1956), pp. 331–355.
4. H. NEIBURGER, *J. Meteorol.*, 6 (1949) 98–104.
5. *Ann. Astrophys. Obs. Smithsonian Inst.*, Vols. 2–6 (1908–1942).
6. S. FRITZ, *J. Meteorol.*, 6 (1949) 277.
7. A. DANJON, *Ann. Obs. Strasbourg*, 3 (1937) 139.
8. K. YA. KONDRATYEV, *Meteorological Investigations by means of Rockets and Satellites* (in Russian), Leningrad (1962), pp. 212–222.

Chapter 7

INSTRUMENTS AND EXPERIMENTAL METHODS

1. INTRODUCTION

We shall assume in this chapter that the reader will be a user rather than a designer or a constructor of instruments, so that he will be more interested in their characteristics and applications rather than in their design and in detailed calculations of the processes involved. The various instruments will, therefore, be described mainly from the point of view of applied physics.

The purpose of the instruments which we shall consider is to measure the energy associated with radiation incident on a plane of given orientation, and to provide information about the spectral and spatial distribution of this energy. This may be done for any component, for example, sky radiation, all the radiation in a particular spectral region, and so on. The instruments will, in general, convert the energy of the incident radiation into another form of energy which can be measured more conveniently. In this connection it is important to keep clearly in view what is intended to be measured, what is actually measured, and how it is measured. This has involved a great deal of discussion; for some publications concerned with this problem, see [1–5].

Depending on the principle on which they are based, the various instruments may be classified as thermal, thermoelectric, photoelectric, and so on. Most instruments are not absolute and require conversion factors so that their readings may be converted into absolute units. Commonly employed units are $cal \cdot cm^{-2} \cdot min^{-1}$, $cal \cdot cm^{-2}$ (which is known as the Langley), $watt \cdot cm^{-2}$, and so on.

A further classification involves the division of the various instruments into standard instruments, secondary standards, and field instruments.

In spite of the great variety of radiation detectors, it is possible to make some general remarks which will apply in particular to instruments which convert radiant energy into heat. This is not the only possible type of conversion: other possibilities will be discussed later.

In instruments converting the incident radiation into heat, a blackened (or partly black and partly white) receiver which is not spectrally selective (in so far as this is possible in practice) is used to provide an indication, or a quantitative reading which is proportional to the incident amount of energy. The energy absorbed by the receiver gives rise to a temperature difference Δt in the receiver, which depends on its thermal properties. The temperature difference in turn gives rise to a heat flow i_h, and there is a thermal resistance R_h to this flow which is analogous to electrical resistance. In fact, in the case of stationary heat flow, a law similar to Ohm's law may be employed, *i.e.*

$$\Delta t = i_h R_h \qquad (7.1)$$

While the temperature difference Δt may be measured with satisfactory accuracy, the evaluation of R_h is very complicated because of heat losses by convection, conduction and radiation, which make an absolute determination almost impossible. The determination of i_h is therefore not a simple matter. However, when R_h can be maintained at a constant value, measurements are possible and may be carried out with satisfactory accuracy. This difficulty may be overcome with the aid of an additional source of energy, for example, by passing a known electrical current creating the same temperature difference Δt at the same or a similar point of the instrument. One example of this procedure will be described in detail when we consider the Ångström compensation pyrheliometer.

A useful procedure which may be employed in the discussion of instruments converting radiation into heat is to draw a 'heat circuit' containing the various thermal elements such as heat capacity, heat resistance and heat flow, by analogy with an electrical circuit containing capacitors, resistors,

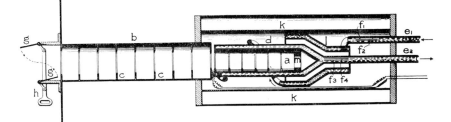

Fig. 7.1. Abbot's continuous-flow standard pyrheliometer.

and so on. The particular form of the heat circuit will depend on the properties of the instrument under consideration. This method of analysis was developed by Courvoisier and Wierzejewski [6,7]. As has already been mentioned, the readings of the instruments must be converted into absolute units, which involves the use of a calibration factor, and the same difficulties arise in the determination of this factor as in the case of R_h.

2. STANDARD INSTRUMENTS

The water-flow pyrheliometer of the Smithsonian Institution

This pyrheliometer was designed by C. G. Abbot at the Smithsonian Institution in Washington in 1905 for the determination of solar radiation in absolute units, and the numerical evaluation of the solar constant by suitable extrapolation [8–10]. This instrument is illustrated in Fig. 7.1 and consists of a cylindrical black-body chamber which completely absorbs the radiation

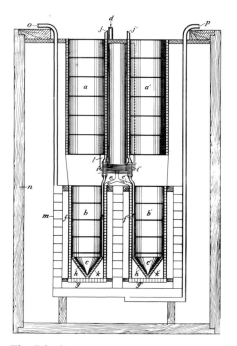

Fig. 7.2. Cross-section of the improved water-flow pyrheliometer.

incident upon it. The chamber has a circular entrance aperture; its walls are thin and have a good thermal conductivity. It is cooled continuously by flowing water, whose temperature is measured before and after it enters the cooling system. When the instrument is oriented so that the radiation is incident normally on the entrance aperture, and if the amount of water, the difference between the above two temperatures, and the time of exposure are all known accurately, then the amount of radiation incident on the entrance aperture may be computed.

In order to achieve accurate results, Abbot used a coil of known resistance to produce the same amount of heat as that produced by the incident radiation. In 1927 Shulgin [11] suggested a modification of the design which was intended to overcome some drift in the readings, and in 1932 Abbot adopted this suggestion. The improvement was achieved by the simultaneous use of two identical black-body chambers, and by dividing the cooling water into two streams. One of the chambers was insolated while the other was heated by an electric current. An adjustment was then made until the heat produced in the two chambers was the same. Provision was made for repeatedly interchanging the role of the two chambers. This modification increased the accuracy of the device and simplified the experimental procedure.

The modified instrument is shown in Fig. 7.2. In this figure a, b, c and a', b', c' are the two identical chambers. Distilled water enters at d and is divided into two streams at e and e'. After flowing round the chambers, the water leaves through j and j'. Thermocouples located at i and i' are used to determine the temperature of the outflowing water. Both chambers are placed in the bath m. Water enters the bath at P and leaves at O; k and k' are heating coils. In order to ensure that the amount of water flowing into the two chambers is the same, they are both insolated and the currents of water are adjusted to produce the same temperature difference. The accuracy of the improved instrument depends neither on the amount of water nor on its temperature, but on the accuracy with which the size of the entrance aperture, the electrical current, and the zero-point of the measuring instrument are known.

The water-stir pyrheliometer

For the purposes of comparison with the water flow instrument described above, Abbot [12] constructed another absolute pyrheliometer in

1912. This was the water-stir pyrheliometer in which a given quantity of water was heated by solar radiation while being continuously stirred.

The Potsdam absolute pyrheliometer

This instrument was first built in 1931 and combines the principles of stirring and compensation by means of an electric current. There are two chambers which can be alternately insolated or heated by the current [13,14].

3. SUB-STANDARD INSTRUMENTS

The geometrical properties of these instruments will be discussed in detail later on. The various standard pyrheliometers are accurate and absolute, but are too complicated and too expensive for frequent use. Sub-standards are therefore used to calibrate and check field instruments. The sub-standards are not intended for absolute measurements or for the determination of the solar constant. They can, however, be calibrated and re-calibrated in absolute units by comparison with standard absolute pyrheliometers. They are capable of providing readings which are proportional to the incident radiant energy, with satisfactory accuracy and reproducibility. They have the further advantage that they can be made portable, and therefore can be used at a number of locations for the purposes of comparison with field instruments. Each instrument has its individual calibration factor which is used to convert its output into absolute units.

The silver disc pyrheliometer

This instrument was designed by Abbot [15] in 1902 who used the preliminary design of Pouillet and Tyndall.

A sketch of the model constructed in 1909 is shown in Fig. 7.3. The silver disc a is 2.8 cm in diameter and 0.7 cm thick. The side of the disc which faces the entrance aperture e is blackened and the bulb of a mercury thermometer b is inserted into the radial hole in the disc. In order to ensure good thermal contact between the thermometer and the disc, a small quantity of mercury which is separated from the disc by a thin layer of steel, is introduced into the radial cavity. The thermometer stem is bent through a right angle as shown. In order to facilitate temperature readings, a slot is cut into the tube supporting and protecting the thermometer which is graduated in steps of 0.1°C between −15°C and +50°C. The silver disc is framed by a cylindrical copper box c which in turn is located in a cylindrical wooden box d. The latter screens the assembly from external influences. The instrument may be accurately pointed at the sun by means of a small hole in the upper support k of the thermometer stem. The solar beam i is made to fall on a mark on the other support. This arrangement for pointing the instrument at the sun is called the diopter. The support k protects the wooden box d from the direct radiation. The solar radiation is admitted into the tube e which contains circular apertures f_1, f_2 and f_3. The last of these apertures has the smallest diameter which is slightly smaller than the diameter of the silver disc. All the inner parts of the device are dull black, and the radiation may be cut off by the shutter g consisting of three nickel-plated discs. The instrument may be made to follow the sun by means of an equatorial mounting.

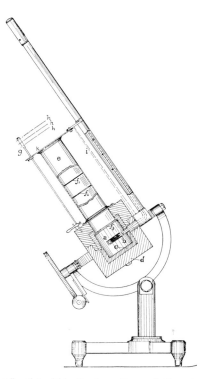

Fig. 7.3. Abbot's silver disc pyrheliometer [16, 17].

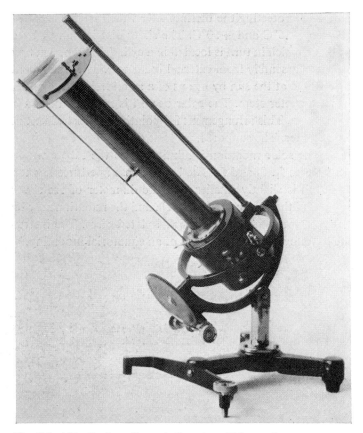

Fig. 7.4. Abbot's silver disc pyrheliometer with long tube.

Two changes were introduced in 1927 into the silver disc pyrhelio-meter [18]. The tube was lengthened and the base enlarged in order to balance the additional weight of the tube (Fig. 7.4). In the original tube each point on the surface of the disc was exposed to a sky cone having an angular diameter of $10°38'$. The total area covered by the tube was approximately 400 times greater than that of the solar disc on the receiver. The original tube length of 15 cm was increased to 32 cm and the exposed area was thus reduced from 0.0043 to 0.0013 of the hemisphere, and the error caused by the haze was reduced from 2.5 % to 0.5 % or even less.

A detailed account of the experimental procedure can be found in the literature [16].

The Ångström compensation pyrheliometer

As mentioned above, the compensation method is capable of providing an accurate value for the heat absorbed in the pyrheliometer in absolute units, and the energy associated with the incident solar radiation can be evaluated from it. An instrument based on this principle was first constructed by K. Ångström [19] in 1893. A slightly modified form of this instrument is still used as a sub-standard. It may also be regarded as a standard instrument. A drawing and a photograph of it are given in Figs. 7.5 and 7.6.

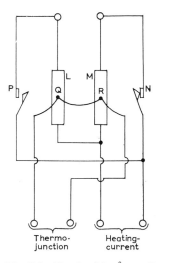

Fig. 7.5. Circuit of the Ångström compensation pyrheliometer [20].

In Fig. 7.5, L and M are two blackened manganin strips which absorb 98 % of the incident radiation. Q and R are two copper–constantan thermocouples electrically insulated from the strips but in thermal contact with them. P and N are keys. In the position shown in Fig. 7.5, strip L is insolated while strip M is shielded from the sun and is electrically heated. The shield is adjustable, so that it can be placed above either of the strips. It is coupled to the electrical circuit, so that as soon as it is placed in position above a particular strip, the latter is automatically heated by an electrical current. It is also possible to expose both strips to the sun with the electrical current switched off. Thermocouples Q and R are connected differentially so that when the thermal resistances of the two strips are equal and the thermocouples are identical, the galvanometer reading, indicating the difference between the two thermal emf's, is zero. In practice, the thermal resistances may not be equal and there may be a non-uniform heat distribution along the strips [21].

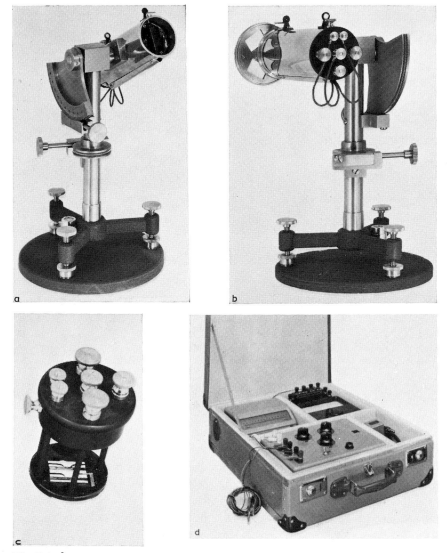

Fig. 7.6. Ångström compensation pyrheliometer [20].
(a) Front view. (b) Rear view with filter holder. (c) Receiver. (d) Auxiliary apparatus.

The radiant energy Q incident on the insolated strip may be computed from the expression

$$Q = Ki_m^2 \tag{7.2}$$

where i_m is the mean value of the electrical current heating the shaded strip and is given by

$$i_m = (i_{L_1} + i_{L_2} + 2i_R)/4 \tag{7.3}$$

where i_{L_1} is the current when the left-hand strip is heated, i_R is the current when the right-hand strip is heated, and i_{L_2} is the current when the left-hand strip is heated again. K is a constant for the particular instrument and has to be determined for each instrument individually. The final expression for Q turns out to be (in cal $\cdot$ cm^{-2} $\cdot$ sec^{-1})

$$Q = Ri_m^2/4.186\ a\beta \tag{7.4}$$

where a is the width of the strips in cm, β is the absorptivity of the strip surfaces, i_m is the current defined above (in amperes) and R is the electrical resistance of the strips per unit length. Since $Q = Ki_m^2$, it follows that the instrumental constant is given by

$$K = R/4.186\ a\beta \tag{7.5}$$

This constant is usually determined by comparison with an instrument for which K may be obtained in absolute units as a result of the determination of a, β and R, each of which can be measured with an accuracy of 0.1% [22].

This type of Ångström pyrheliometer is found to be inaccurate because the diaphragm in front of the insolated strip shades approximately 0.15–0.20 cm of each end, whereas the electrical current heats the strip along its entire length [20, 23, 24]. Less current is therefore needed to heat the strip in order to compensate for the heat supplied by solar radiation. Recent studies have shown that a 2% correction must be added to K for all the Ångström pyrheliometers constructed before January 1, 1957. A correction of 1.5% must be added in the case of instruments issued after this date.

In the latest Ångström pyrheliometers the shaded ends extend over less than 0.02 cm and a correction is therefore unnecessary. These latest instruments have different geometrical dimensions. A comparison between an early type and a recent type is given in Table 7.1 (cf. Fig. 7.8).

Another error of this instrument is due to heat convection from the

TABLE 7.1. Comparison between an early type (No. 158) and a recent type (No. 168) of the Ångström pyrheliometer.

	Section along the strips		Section across the strips	
	No. 158	No. 168	No. 158	No. 168
R (mm)	10.0	10.0	2.0	2.5
r (mm)	8.8	9.5	1.0	1.0
h (mm)	43.0	80.0	43.0	80.0
α_1 (°)	12.6	7.1	2.6	7.1
α_2 (°)	1.6	0.4	1.3	1.1
α_3 (°)	23.1	13.7	3.9	2.5

edges of the strip and is known as the edge effect. The latter affects the end correction mentioned above. A method of computing this error has been described by Ångström [20].

The auxiliary apparatus used with the Ångström pyrheliometer is of great importance because the reproducibility and accuracy of the measurements depend upon it. It includes a sensitive galvanometer, which is used to determine the temperature difference between the two strips and must be capable of indicating a difference of 0.01°C. A galvanometer having an internal resistance of 10 ohms and a sensitivity of $10^{-8} - 5 \times 10^{-8}$ amperes per scale division may be used (cf. Fig. 7.6). The auxiliary apparatus must also include a milliammeter which is used to measure the current i_m in Eq. (7.3). Since $Q = Ki_m^2$, it follows that $dQ/Q = 2 \ di_m/i_m$ and hence the accuracy to which i_m is measured should be twice that of Q. The milliammeter must be shaded in order to avoid further errors.

Other details of the Ångström pyrheliometer and instructions for its use are given in Instruction Manual No. 2 issued by the Swedish Meteorological and Hydrological Institute, Instrumental Division (Stockholm, 1957).

The Smithsonian version of the Ångström pyrheliometer

Four modifications of Ångström's original instrument were introduced at the Smithsonian Institution: (1) The instrument was enclosed in a wooden box in order to prevent effects due to external temperature changes and wind. (2) All electrical contacts were soldered. (3) A mercury switch was provided for reversing the heating current automatically when the shutter is reversed. (4) A platinum-coated cone was inserted into the base of the tube; this cone reflects the radiation below the strip in the direction of the tube

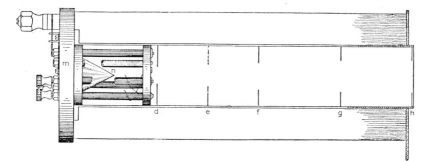

Fig. 7.7. Smithsonian version of the Ångström compensation pyrheliometer (d, e, f, g and h are diaphragms).

walls, thus preventing the heating of the strip by this radiation [25]. The instrument is illustrated schematically in Fig. 7.7.

4. PYRHELIOMETRIC SCALES

Abbot's silver disc pyrheliometer and the Ångström compensation pyrheliometer are sub-standards which are generally accepted for the comparison and recalibration of solar-radiation measuring instruments. A 'pyrheliometric scale' is associated with each of them. The first scale was adopted in 1905 by the International Meteorological Organisation (IMO), now known as the World Meteorological Organisation (WMO), and was based on the Ångström pyrheliometer. In 1913 another scale, based on the results obtained with the water-flow pyrheliometer, was introduced by Abbot. This scale is known as the 'revised Smithsonian pyrheliometric scale' and was used for calibrating the silver disc pyrheliometer. The two scales are not identical. As was noted above, there are differences in the apertures, shapes and dimensions, and in the fraction of the hemisphere subtended by each instrument. In 1932 the Smithsonian Institution announced that its 1913 scale was too high by 2.5%. This announcement was repeated in 1934, 1947 and 1952. Despite this, the Smithsonian Institution continued to calibrate its pyrheliometers against the 1913 scale in order to preserve continuity.

In spite of the fact that the two scales were frequently compared (with the sun as a source), no clear results were obtained. The instruments were not comparable for reasons mentioned above, but the difficulty was partly obviated by the use of lamp sources under laboratory conditions. However, this involves other difficulties; for example, it was almost impossible to obtain a uniform irradiation of the receiver surface, owing to the large receiver area. Extensive tests were carried out to check the constancy of the two scales and to find a numerical conversion factor [20, 26–31].

The Radiation Commission of the IAM, and the Working Group on Radiation of the WMO, recommended a new scale at the joint International Radiation Conference at Davos (Switzerland) in September 1956. The new scale is known as the International Pyrheliometric Scale 1956 (IPS 1956) and was based on the two former scales mentioned above. Measurements made on the original Ångström scale must now be increased by 1.5%. The difference between the 1913 Smithsonian scale and original uncorrected Ångström scale was 3.5%, so that measurements based on the 1913 Smithsonian scale must be reduced by 2% [26, 30].

5. INSTRUMENTS FOR THE MEASUREMENT OF DIRECT SOLAR RADIATION: PYRHELIOMETERS

The function of these instruments is to measure the radiation of the sun only, along a surface perpendicular to the solar beam. It is practically impossible to achieve this object, because even a small change in the direction of an instrument with a small aperture adapted to the disc of the sun gives rise to large errors. The instruments are therefore constructed so as to include a part of the circumsolar radiation. In such instruments the receiver has to be protected from scattered radiation (sky radiation) and is usually located at the bottom of a tube. For accurate aiming at the sun, a diopter giving a narrow solar beam parallel to the central axis of the tube is provided. The dimensions and shapes of the receivers and tubes vary, so that different amounts of circumsolar radiation are included in measurements carried out with different instruments. Since the geometry of the tubes is of considerable importance, the pyrheliometers will be considered from this standpoint, as well as from that of the receiver. To begin with, let us discuss some general points.

Pyrheliometers with circular receivers and cylindrical tubes

The geometrical features of a pyrheliometer with a circular opening and receiver are illustrated in Fig. 7.8 in which α_1 is the angular aperture, α_2

is the slope angle, α_3 is the limiting angle, α_4 is the covering angle, h is the length of the tube, b is a generator line, l is the length of a diagonal of the tube, R is the radius of the upper opening, and r is the radius of the lower opening.

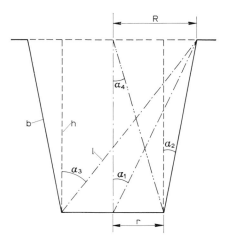

Fig. 7.8. Geometry of a pyrheliometer tube.

Although the tubes are cylindrical, the sketch shown in Fig. 7.8 is correct since there are diaphragms inside the tube, with diameters decreasing towards the receiver [32] (as in the case of the silver disc pyrheliometer described above, Fig. 7.3).

The determination of the quantity of solar radiation falling on the receiver is complicated by a number of factors. Thus, neither the radiation of the solar disc nor the circumsolar radiation are uniform, and the receiver is not uniformly sensitive.

Linke used the parameters $a = R/r$ and $b = h/r$, where R and r are the radii of the entrance aperture and of the receiver respectively, and h is the length of the tube, to describe the geometrical features of a pyrheliometer. The geometry of the pyrheliometer determines the ratio of the quantity of radiation Q entering the tube to the quantity Q_1 falling on the receiver. When the irradiation is uniform and the entrance aperture and the receiver are circular, concentric, and parallel, the above ratio may be evaluated from

$$Q_1 = Q \int_A \int_{A'} \frac{\cos \beta \, dA \, dA'}{d^2} \qquad (7.6)$$

where dA and dA' are the surface elements of the entrance aperture and receiver respectively, β is the angle between the normals to these two elements, and d is the distance between them. According to Lambert, the result of the integration is

$$Q_1 = \tfrac{1}{4}\pi^2 Q(1-b)^2 \tag{7.7}$$

while according to Weickman, who used certain approximations (cf. Eq. (7.9)), the result is

$$Q_1 = \pi^2 Q h^2 \sin^2 \alpha_1 \sin^2 \alpha_4 \tag{7.8}$$

Pyrheliometers with rectangular entrance apertures and receivers

For uniform irradiation, and with the entrance aperture and receiver areas arranged as above, the above integral is quite difficult to evaluate and contains the dimensions of the two rectangular areas and the perpendicular distance between them. Using the same approximations as above, Weickman has shown that Q_1 is given by

$$Q_1 = 16\, Q h^2 \sin \alpha_1 \sin \alpha_1' \sin \alpha_4 \sin \alpha_4' \tag{7.9}$$

where α_1, α_1', α_4 and α_4' are the angular apertures corresponding to the two sides of the rectangle for the entrance aperture and the receiver, respectively.

Distribution of radiation on the receiver surface

Eqs. (7.8) and (7.9) give the total amount of energy reaching the receiver when every point on its surface is irradiated by all points on the entrance aperture. This results in a certain distribution of energy on the receiver. It will become evident from the description of the various receivers that a knowledge of this distribution is important in practice.

In the case of a uniformly insolated entrance aperture, the distribution of energy on the receiver surface will be a function only of the distance x from the centre of the receiver. Calculations show that the energy reaching an annular surface element dA_1 of the receiver per unit time is given by

$$dQ_1 = \tfrac{1}{2}\pi Q \left[1 - \frac{h^2+x^2-r^2}{((h^2+x^2-r^2)^2+4r^2h^2)^{\frac{1}{2}}} \right] dA_1 \tag{7.10}$$

where x is the mean radius of the annular element.

The assumption of uniform irradiation at the entrance aperture is not valid for reasons explained above. Moreover, the receiver surfaces may not be of uniform sensitivity. An exact computation is therefore rather difficult and cannot be given here [33,34].

Frequently employed pyrheliometers

Now that the geometry and constructional properties of pyrheliometers have been briefly dealt with, some frequently employed instruments will be described in greater detail and their properties evaluated. In general, for all pyrheliometers the following criteria will be considered:

(1) Absence of zero drift.

(2) The calibration factor must be independent of temperature.

(3) The output must be independent of the wavelength for a given radiation intensity.

(4) The calibration factor must be independent of the intensity.

(5) Minimum time lag between the beginning of the exposure and the maximum recorded output.

(6) The calibration factor must be independent of time.

(7) The instrument must be protected against extraneous effects.

Michelson's bimetallic pyrheliometer

The first instrument incorporating the bimetallic strip, which bends as a result of heating by solar radiation, was built in 1905 by Michelson in Russia [7, 35–37]. The strip was made of iron and nickel steel, 1.1 cm long, 0.17 cm wide and 0.0005 cm thick, and was covered with platinum black. Because of the small heat capacity and the difference in the expansion coefficients, the instrument was fairly sensitive.

It can be shown that a temperature change Δt will give rise to a radius of curvature R given by

$$R = d/(\alpha_1 - \alpha_2)\Delta t$$

where $\alpha_1 - \alpha_2$ is the difference in the linear thermal expansion coefficients of the two metals and $2d$ is the total thickness of the strip.

A device of this type is illustrated in Fig. 7.9 in which 1 is the bimetallic strip, 2 is a quartz fibre 2–8 μ in diameter, 3 is an aluminium tube (0.2 cm wide, 0.1 cm high; wall thickness 9 μ), and 4 is a plane mirror. The

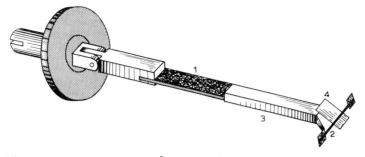

Fig. 7.9. Michelson's bimetallic strip pyrheliometer.

bending of the strip is observed by means of a small telescope. The instrument is not temperature-compensated and hence the zero point depends on the ambient temperature, in spite of protection from extraneous effects. This original design is now seldom used and will not be discussed in detail.

Improved Michelson pyrheliometers

An improved version, known as the Michelson–Marten type, is used at some solar stations [35]. The bimetallic strip consists of constantan–invar combination (1 cm long, 0.2 cm wide and 0.0005 cm thick). The dimensions of the entrance aperture are not adapted to those of the receiver; they are 0.8 cm × 0.35 cm with entrance angles of 9.5° and 4° respectively.

Let us consider this model, in terms of the criteria listed above:

(1) Since there is no thermal compensation, the zero point depends on the temperature and must be established in each case by cutting off the incoming radiation.

(2) The conversion factor is temperature-dependent. A special screw for correcting this is included, and an appropriate graph is provided with each instrument. An example of the temperature dependence of the conversion factor is given below:

t (°C)	−20	0	+20	+40
C.F.	0.0243	0.0250	0.0257	0.0265

Fig. 7.10 shows a typical correction curve.

(3) Nothing is known about the spectral selectivity; by using appropriate blackening this effect may be excluded.

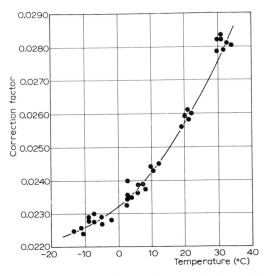

Fig. 7.10. Example of a correction curve for the calibration factor of a Michelson pyrheliometer [38].

(4) Nothing is known about the dependence of the calibration factor on the intensity.

(5) The time lag is between 20 and 30 seconds.

(6) Provided there is no mechanical damage the conversion factor is independent of time.

(7) External effects can be excluded by exposing the instrument for some minutes to outside conditions prior to the actual measurements. The instrument is provided with an envelope which protects it against sudden temperature changes and wind velocity.

Further improvements of the Michelson pyrheliometer were introduced by Büttner. A second bimetallic strip was placed below the first in order to compensate for the bending due to changes in the ambient temperature. The upper strip is irradiated and shades the lower strip. An almost complete compensation is possible. This modification gives rise to an increase in the accuracy and a reduction in the necessary time of exposure. The aperture of the new instrument, which subtends an angle of 3°, is covered by a silica window with the result that the output is reduced by 7.4%. This must be added to the final result. It is found that the inclination of the instrument affects the accuracy, by an amount not exceeding 1%. As regards

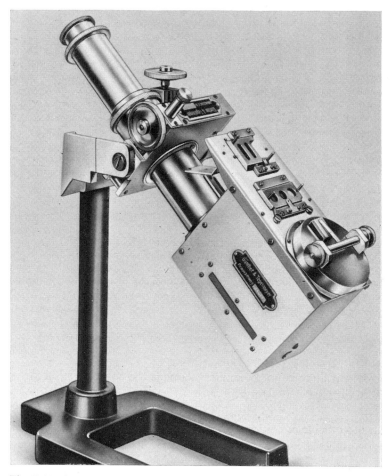

Fig. 7.11. A recent Michelson–Büttner pyrheliometer (Courtesy Günther und Tegetmeyer G.m.b.H., Frankfurt/M., Germany).

criteria 2–7, this instrument is identical with that described above. A photograph of it is shown in Fig. 7.11.

The Armour–Linke–Feussner pyrheliometer

The Armour pyrheliometer which is based on the Linke–Feussner design is shown in Fig. 7.12. It is so designed that it can be used either for direct solar radiation, or for measurements on sky radiation over a limited part of the sky. It can be used with the internationally adopted filters which will be considered later in this chapter. The receiver is a Moll type thermopile which is sensitive up to 40,000 Å. The theory of this receiver is given by Bener [54]. It can be used with short- and long-wave radiation, the separation being achieved by means of a silica window. The thermopile consists of 40 constantan–manganin strips, arranged in two groups of 20. The diameter of the thermopile is 10 mm, the output is 10 mV per cal $\cdot$ cm^{-2} $\cdot$ min^{-1}, and the resistance is about 60 ohms. The tube contains massive copper rings which serve as diaphragms with apertures decreasing in diameter towards the thermopile. The motion of air inside the tube is prevented by special shaping of the diaphragms.

The instrument has the following characteristics on the basis of the seven criteria mentioned above:

(1) There is no detectable zero drift.

(2) The temperature dependence of the calibration factor for direct radiation is of the form $1 + \alpha(20° - t°)$, where $\alpha = 0.002$.

(3) The readings of the instrument are independent of wavelength.

(4) The dependence of the calibration factor on the intensity has not been adequately investigated.

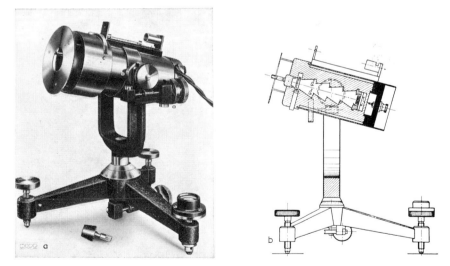

Fig. 7.12. Left: Armour pyrheliometer (after Linke–Feussner). Right: Cross-section of the Linke–Feussner pyrheliometer. (Courtesy Kipp & Zonen, Delft, The Netherlands.)

(5) The time of response is 8 seconds for maximum output. A complete series of measurements, including the determination of the zero point, the total output, the output for three different filters and a final check of the zero point can be completed in one minute.

(6) The dependence of the calibration factor on time has not been investigated in detail.

(7) The receiver consists of connected thermopiles one of which is irradiated and the other is shaded. Since both are exposed to identical extraneous influences, a minimum of interference due to sudden temperature changes, wind, and so on, is achieved. This, together with the large mass of the instrument, ensures that extraneous influences are virtually negligible.

The Moll–Gorczynski pyrheliometer

This pyrheliometer incorporates a Moll type thermopile (cf. Fig. 7.19) and was designed by Gorczynski [35,39]. Scattered radiation is excluded by a number of diaphragms and an equatorial mounting is used to follow the sun. The thermopile consists of 14 constantan–manganin strips, 0.85 mm wide, 10.5 mm long and 0.007 mm thick. The total area is 10.5 mm × 12.4 mm. The resistance is usually between 8 and 10 ohms, and the output signal is 22 mV per cal $\cdot$ cm^{-2} $\cdot$ min^{-1}.

An examination of this instrument in the light of the above criteria yields the following results:

(1) No exact information is available about zero drift.

(2) The sensitivity is found to decrease by 0.2 % per 1°C.

(3) The output of the device is independent of wavelength in the range 2000–100,000 Å; for longer wavelengths this is not certain.

(4) Nothing is known about the dependence of the calibration factor on the intensity.

(5) The response time is 8–10 seconds.

(6) Ageing effects were found to be observable only immediately after the construction of the device.

(7) A silica window is the only protection against extraneous influences.

The Eppley pyrheliometer

The instruments considered so far incorporated a thermopile as the receiver. We shall now consider pyrheliometers in which the receiver is made of a metal whose surface is partly blackened and partly white, although in some cases it may be completely black. The black and white parts are insulated from each other and thermocouples are arranged underneath the receiver surface so that junctions in thermal contact with the black part of the receiver are the hot junctions while those in thermal contact with the white part of the receiver surface are the cold junctions.

Let us consider the composite receiver (partly black and partly white). Since the two parts of the receiver surface have different absorptivities, a temperature difference between them will be established when they are exposed to radiation. This will in turn give rise to a thermal emf at the thermocouple output.

An example of this type is the Eppley pyrheliometer which is illustrated in Fig. 7.13. It incorporates a circular silver receiver which is coated with lamp black. The receiver is 7/32″ in diameter and 0.0001″ thick. The thermocouple is of the eight-junction copper–constantan type. The emf produced is approximately 2 mV per $cal \cdot cm^{-2} \cdot min^{-1}$ and the internal resistance is approximately 6 ohms. The receiver is located at the bottom of a cylindrical brass tube carrying a set of diaphragms.

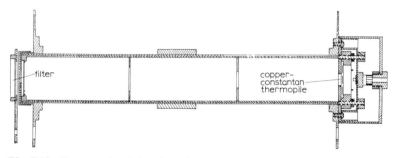

Fig. 7.13. Cross-sectional drawing of the older type of Eppley normal-incidence pyrheliometer.

The angle subtended by any point on the receiver is 5°41′30″. The total length from the receiver surface to the inner surface of the silica window closing the tube is 10 times the diameter of the first diaphragm. A diopter is used to determine the direction of the sun. The inner wall of the tube is blackened and contains dry air at atmospheric pressure. The tube is sealed with the silica window which is 1 mm thick.

An examination of the properties of this instrument on the basis of the seven criteria mentioned above gives the following results:

(1) There is no zero drift.

(2) The change in the calibration factor is less than 0.2% for an ambient temperature rise of 1°C.

(3) The output is independent of wavelength.

(4) The intensity dependence of the calibration factor is unknown.

(5) 98% of the maximum reading is obtained in 5 seconds.

(6) No ageing was observed after a number of years.

(7) The effect of extraneous influences is negligible.

A special frame is provided just outside the silica window for the insertion of filters. In continuous operation the instrument should be mounted on an equatorial drive.

A brief description of this instrument is given in Bulletin No. 9 issued by the Eppley Laboratories of Newport, R.I., U.S.A. (November 1956).

A more recent model of the Eppley pyrheliometer incorporates various improvements and changes (Fig. 7.14). The receiver is in the form of a thin silver disc (9 mm in diameter) which is coated with Parson's optical black lacquer. Fifteen junctions of fine bismuth–silver thermocouples are in thermal contact with, but electrically insulated from, the lower surface of the disc. The cold junctions are in thermal contact with the copper tube of the instrument. A rotatable frame for three filters is provided. The silica window is removable. The output is 3–3.5 mV per $cal \cdot cm^{-2} \cdot min^{-1}$ and the internal resistance is about 400 ohms at 25°C, including the thermistor circuit.

An examination of the properties of this instrument in the light of the above given criteria gives the following results:

(1) No zero drift.

(2) 1% accuracy is achieved between $-40°$ and $+50°C$ by a special temperature compensator.

(3) Readings are independent of wavelength in the range 3000–30,000 Å.

(4) The intensity dependence of the calibration factor is unknown.

(5) 98% of the maximum output is reached in 20 seconds.

(6) Ageing properties are unknown.

(7) Extraneous influences are negligible.

The Savinov–Yanishevskiy pyrheliometer

In this instrument (Fig. 7.15) the receiver is also in the form of a silver

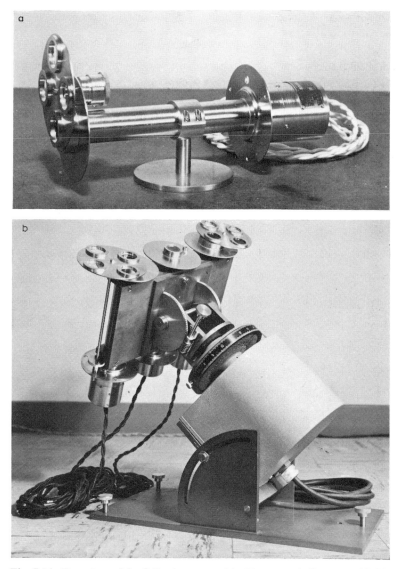

Fig. 7.14. Recent model of Eppley normal-incidence pyrheliometer. (a) Unmounted. (b) Mounted on an equatorial mounting.

disc (1) which is 11 mm in diameter and 0.003 mm thick. The central part of the disc, which is 3.5 mm in diameter, is removed. The surface facing the sun is coated with a black mixture of soot and amber. The other surface is covered with a sheet of thin cigarette paper attached to the disc with the aid of shellac. Alternate junctions of the thermocouple are attached to this paper (2 and 5), while the remaining junctions (3) are attached to a copper ring (4) which is pressed into the tube of the instrument. The aperture of the tube is 5°.

The thermopile consists of 36 pairs of manganin–constantan strips, $6.0 \times 0.3 \times 0.04$ mm. The output leads of the thermocouple are shown (6). The output signal is 4–7 mV per cal $\cdot$ cm^{-2} $\cdot$ min^{-1} at an internal resistance of 13–20 ohms.

In the cross-sectional drawing 1,2 is the part of the copper ring 4 which is covered by the thermopile and 3,5,6 are the ends of the thermopile.

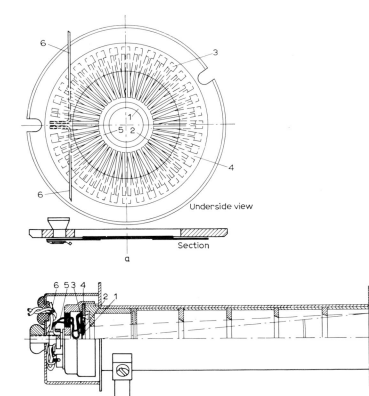

Fig. 7.15. Savinov–Yanishevskiy pyrheliometer. Top: receiver. Bottom: cross-sectional drawing.

An examination of the device from the point of view of the above seven criteria gives the following result:

(1) Zero drift: not known.
(2) The overall temperature dependence is 0.02 % per 1°C.
(3) Dependence of the output on the wavelength: not known.
(4) Dependence of the calibration factor on the intensity: not known.
(5) Maximum output is obtained in 14–25 seconds.
(6) Dependence of the calibration factor on time: not known.
(7) No information is available about extraneous effects.

Auxiliary apparatus used with pyrheliometers

Pyrheliometers can be used either for measuring the instantaneous values of the intensity or for continuous determinations. In the former case the pyrheliometer is mounted on a stand which can be rotated so as to follow the sun and the output can be measured with a suitable millivoltmeter, although a potentiometer is preferable. The Michelson pyrheliometer is self-contained and needs no auxiliary measuring instrument. In continuous operation the pyrheliometer has to be mounted on an equatorial drive so that it constantly faces the sun. The output is then recorded by a millivoltmeter or a recording potentiometer, preferably with automatic integration.

6. INSTRUMENTS FOR THE MEASUREMENT OF GLOBAL RADIATION: PYRANOMETERS AND PYRANOGRAPHS

These instruments are used to determine the solar and sky radiation together on a horizontal plane, *i.e.* global radiation (G). The receiving elements must be horizontal and freely exposed to the celestial hemisphere, but must be protected from back radiation from the ground and the surroundings.

We shall examine the various forms of these devices on the basis of the following criteria:

(1) Cosine response, *i.e.* the extent to which the equation $G = I \cos z_{\odot} + D$ is valid, where $z_{\odot}$ is the zenith distance of the sun and D is the intensity of the scattered radiation (sky radiation). This response is subject to errors due to the variation in the absorption by the receiver surface with the angle of incidence, unevenness in the receiver surface and in the optics of the covering glasses, incorrect levelling of the receiver surface, and variation of

the receiver temperature with the elevation of the sun. The first three errors are related to the instrument and cannot be corrected, but the other two can be corrected by proper levelling and by taking the temperature coefficient into account. The calibration factor may show a diurnal and an annual cycle.

(2) The temperature coefficient of the calibration factor.

(3) Changes in the output with time.

(4) The absence of spectral selectivity.

(5) Time of response.

(6) Absence of zero-point depression.

(7) Dependence of the calibration factor on the quantity of irradiated energy.

(8) Effect of extraneous influences such as temperature changes, wind, etc.

The Robitzsch pyranometer

The theory of this instrument, which is also known as the Robitzsch actinometer, has been given by Courvoisier [6, 7]. The results may be summarized as follows [40]. The computed calibration factor differs from the experimental result by only a few percent. The temperature coefficient of the calibration factor is about 0.8% per 1°C. The temperature coefficient gives rise to considerable changes in the sensitivity of the device, and this must be taken into account in the evaluation of diurnal and seasonal records. The time necessary to attain 98% of the maximum output is 8 minutes. All other criteria will be dealt with later on.

The instrument designed originally by Robitzsch in 1915 is similar to a thermograph. A blackened bimetallic strip, fixed at one end and free at the other, is irradiated. The strip bends as a result of temperature changes and the motion of the free end is transferred mechanically to a pen which records the displacement on a rotating drum. Various complications are involved in measuring the amount of solar radiation, *e.g.* special precautions must be taken to exclude the influence of the ambient temperature on the strip.

There are now four types in use. Two are manufactured by R. Fuess (Berlin–Steglitz, Germany), one by C.F. Casella (London) and one by SIAP (Società Italiana Apparecchi Precisione, Bologna, Italy). These four types will now be described and considered with respect to the eight basic criteria.

The Fuess pyranometer

The early instruments manufactured by Fuess are very seldom used at present and will not be described. The three-strip type 58C, which is extensively used, will now be considered. The instrument is very similar in construction to the Casella instrument illustrated in Fig. 7.17, except for the white plate below the receiver. The receiver consists of three bimetallic strips, the central one being black and the outer strips white. The three strips are attached to a small metal rod (on the left of the picture) which is thermally insulated from them. On the right the white strips are fixed and the black strip is free to move. The bending of the black strip is transferred to a pen by means of levers. A white metal plate, placed below the receiver (not shown in Fig. 7.17), protects the housing and the moving mechanism from solar radiation. The instrument is enclosed in a metal case with flat glass windows in front, and a hemispherical glass cover above the receiver. The instrument is oriented so as to face north through one window with the strips lying along the east–west direction.

It is convenient at this point to consider a certain confusion which has arisen with regard to the calibration factor. It has been found that different instruments indicate quantities of radiation which are too high by 50% or more. This is not due to the design of the instruments [41] but to the varying astronomical and atmospheric conditions. Many effects, such as absorption by the glass cover or by the coatings on the receiver strips, influence the calibration factor and may be included in it provided they remain constant.

As regards the above eight basic criteria, the properties of the instruments may be summarized as follows:

(1) The cosine response depends on the elevation of the sun $\gamma_\odot$ and the sun's azimuth $A_\odot$. The effect of changes in $\gamma_\odot$ on the calibration factor increases with the increasing $\gamma_\odot$ and this gives rise to a decrease in the sensitivity, *e.g.* when $\gamma_\odot$ changes from $10°$ to $70°$, the calibration factor will increase by about 40% [42]. The effect of $A_\odot$ on the calibration factor can be determined in two ways, namely, by installing two similar instruments, one in the regular position and the other with its window protected from direct solar radiation (at right angles to the first), or by taking measurements for different $A_\odot$ at constant $\gamma_\odot$. The first method shows a difference of around 2% while the second method yields a somewhat larger figure [43]. The calibration factor exhibits a diurnal and an annual cycle. Between March and September, it is generally larger in the afternoon and in the early morning.

There is a trend for the calibration factor to increase with increasing $\gamma_\odot$. The annual cycle exhibits the following features. Between November and January the calibration factor is higher in the morning than in the afternoon and there is a decrease in the calibration factor with increasing $\gamma_\odot$. The mean monthly values of the calibration factor reach a maximum in the summer and a minimum in the winter [44]. The optical properties of the hemispherical glass cover are found to affect the calibration factor with varying $\gamma_\odot$ and $A_\odot$, through the lens effect and the motion of the caustic. If, however, the glass cover is polished correctly and is thin enough, and if its dimensions are suitably matched to those of the receiver, there will be no optical effect [45].

(2) It has been found experimentally that a 20% change in the calibration factor occurs when the temperature is increased from 0° to 25°C, and there is also a difference between the morning and the afternoon [46].

(3) The age of the instrument will only have a significant effect if changes occur in the coatings of the strips. The calibration factor increases and the sensitivity decreases with time as the black coating becomes grey, then white-yellow, and finally cracked, so that the metal base is exposed. These effects appear after two or three years [42].

(4) The calibration factor includes contributions due to effects associated with the glass cover and the dependence of the absorptivity of the receiver coatings on wavelength.

(5) No exact information about the time of response is available.

(6) The constancy of the zero point is of particular importance because the amount of energy insolated during a given period of time is determined by integration between the curve and the 'zero line'. The constancy of the zero point, and its depression in the absence of the solar radiation, depend on the heat transfer between the receiver and the housing, between the housing and the surroundings, and so on. This is a very complicated process and only a few examples of it will be considered here. The black and white strips are at different temperatures and the heat transfer between them gives rise to a lower sensitivity. The white strips are nearer to the glass covers so that they are more affected by cooling than the central strip. The temperature of the housing has a considerable influence on the temperature of the receiver.

(7) The calibration factor has been found to increase with the amount of insolated energy. This increase may be as high as 20% when the intensity changes from zero to 1.5 cal $\cdot$ cm^{-2} $\cdot$ min^{-1} [43]. In view of this change the readings of the instruments are not proportional to the insolated energy and

this must be taken into account in the analysis of the readings. The calibration factor is found to increase with the ratio of the insolated energy to the reading of the instrument [43].

(8) When the instrument is tightly closed, extraneous influences are confined to the heating or cooling of the metal housing by wind.

It is clear from the above discussion that the same calibration factor cannot be used for all radiation values or for all seasons. Each instrument must be calibrated by comparison with a standard instrument.

The recent Fuess model 58D is an improved version of the 58C. In this instrument the receiver consists of three parallel blackened bimetallic strips which are 85 mm long and 15 mm wide. Three other strips are arranged below them in order to compensate for the effect of the ambient temperature. Complete compensation under steady-state conditions is achieved by means of an additional bimetallic strip in the transmission mechanism between the receiver and the recording pen. Fig. 7.16 shows a photograph of the new model (the receiver is not actually as illustrated here; all upper surfaces of the strips are black).

Fig. 7.16. Fuess model 58D of the Robitzsch pyranometer.

The instrument is divided into two parts. The upper part is cylindrical and contains the receiver. Because of its shape, the instrument is not very sensitive to changes in the azimuth or in the wind direction. The second part, *i.e.* the box contains the drum bearing the chart on which the readings are recorded. No manipulations in the neighbourhood of the receiver are necessary and the chart can only be changed by opening the box. Because there are now three strips instead of one, as in the older model, the width of the chart can be increased from 50 mm to 80 mm. Friction between the pen and the chart, which gives rise to errors at small amplitudes, is reduced by means of a small vibrator incorporated in the instrument. Provision is made for introducing time markers on the chart and for zero adjustment.

The amount of experimental data available at present is not sufficient to evaluate the instrument from the point of view of criteria 1, 3 and 4. The influence of the lens effect, of the caustic curve effect, and of the housing is also unknown. All these factors are included in the correction factor F, which is unity at $\gamma_\odot = 60°$ but is in general an unknown function of $\gamma_\odot$.

The glass cover is an optically ground and polished hemisphere which is transparent to all wavelengths except for the ultraviolet and infrared with wavelengths in excess of 20,000 Å.

The Casella pyranometer

This instrument was designed by the Meteorological Research Committee of the Meteorological Office (London) as a result of an extensive study of the Fuess pyranometer. The white plate below the receiver was rejected because it was shown that the back radiation from this plate gave rise to changes in the sensitivity of up to 30% between $\gamma_\odot = 15°$ and 35°. Moreover, the white strips were shielded by thin aluminium sheets painted white and placed lower than the black strips. In order to increase the sensitivity and reduce the time constant, the material of the strips was changed from invar–brass to 'morflex' which is stiff and gives a high specific bending. This modified form of the Fuess–Robitzsch pyranometer is referred to as M.K.II. Laboratory tests carried out over a period of six months showed that the calibration factor remained constant to within 2–3%. It was found that the influence of $\gamma_\odot$ and $A_\odot$ was such that with the light source in the south, the sensitivity showed the smallest increase as the source was moved to east or west. The reason for this was that in this position some radiation penetrated between the strips and reached the white surfaces.

A further model called M.K.III, which is illustrated in Fig. 7.17, was designed on the basis of the results obtained with M.K.II and is manufactured by Casella of London. The main features are as follows.

Two additional bimetallic strips are introduced in order to reduce the

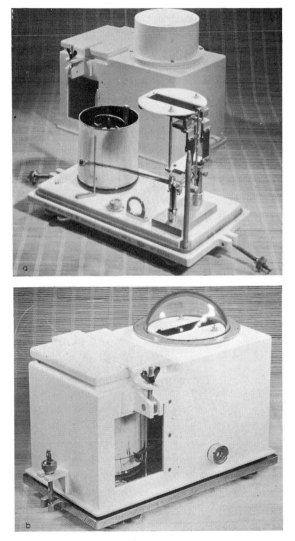

Fig. 7.17. Casella model of the Robitzsch pyranometer. Top: open. Bottom: covered.

east–west asymmetry. Their lengths are one-half of the length of the strips in the receiver and they are mounted vertically in the support (these are the two black vertical strips on the right hand side of the picture of the open instrument).

This arrangement is designed to compensate for any departure of the receiver from flatness which may be due to changes in the ambient temperature. The receiver can be adjusted for any ambient temperature by means of two U-pieces and screws. In addition, the black strip was made wider and the two shaded strips narrower. It was found that a width-to-length ratio of 1:4 eliminated marked transverse bending. The shaded strips were protected from solar radiation for all values of $\gamma_\odot$ and $A_\odot$. The instrument is provided with calibration charts from which the amount of solar radiation incident in a given time can be determined in absolute units, being the area on the chart enclosed between the curves and the zero line. For accurate measurements the zero point must be determined precisely. This is done with the aid of an effective zero which is a line parallel to the horizontal axis of the chart, drawn through the highest point of the trace reached during the night. The effective zero is different from the true zero because the receiver loses heat during the night by dark radiation. The true zero is determined by placing a metal cap over the glass hemisphere.

From the point of view of the criteria listed above, the M.K.III instrument exhibits the following properties [47]:

(1) The cosine response for different values of $\gamma_\odot$ and $A_\odot$ is very nearly exact.

(2) Temperature coefficient of the calibration factor: not known.

(3) The ageing of the coatings and of the glass cover have not been adequately investigated.

(4) Spectral selectivity: not known.

(5) In the absence of the friction between the pen and the chart, the response time is approximately 4 minutes but may be increased to about 7 minutes as a result of friction.

(6) The zero drift was discussed above.

(7) The linear part of the characteristic is above 1.33 cal $\cdot$ cm^{-2} $\cdot$ min^{-1}.

(8) External influences were discussed above (see zero point).

It must be stressed that the M.K.III will only exhibit the above characteristics if all the parts are properly made. This is particularly important as regards the pivots, the flatness of the strips, and so on.

The SIAP pyranometer

This instrument, which is also known as the S.O.28, is illustrated in Fig. 7.18. The receiver consists of three nickel–iron strips. The central strip is blackened and the others are covered by a highly reflecting coating. The overall accuracy is ± 7.0 % as compared with the standard instruments. The glass sphere covering the receiver cuts off all radiation with wavelengths

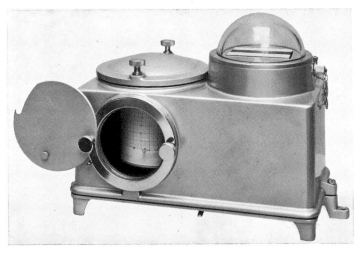

Fig. 7.18. SIAP model of the Robitzsch pyranometer.

greater than 20,000 Å. Tests reported by Courvoisier [48] gave the following results: (a) losses in the transmission system give rise to a 5 % scatter in the calibration points, (b) the temperature compensation is insufficient and considerable errors may arise due to this, (c) the dependence of the calibration factor on the temperature is given by $K_t = 19.4 \, (1 + 0.0033 \, t_A)$.

Other pyranometers and pyranographs of similar design are described elsewhere [49–53].

The Moll–Gorczynski pyranometer

In this instrument [6,7] the receiver is a Moll thermopile consisting of fourteen constantan–manganin strips 10 mm long, 1 mm wide and 0.005 mm thick. The material is sufficiently rigid so that thin strips can be used. The

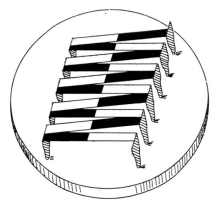

Fig. 7.19. Moll thermopile.

fourteen strips are arranged in a zig-zag pattern, as shown in Fig. 7.19, forming approximately a square of 10 mm × 14 mm. The strips are in thermal contact with, but are electrically insulated from, the copper plate, which has a large thermal capacity. A black varnish of low thermal conductivity fills the gaps between the strips, forming a flat surface which is maintained in a horizontal position. Owing to the poor thermal contact between the strips, each of them can be treated separately as far as heat transfer is concerned. The heat absorbed by the receiver from the solar radiation is transferred to the air by convection, to the copper plate by conduction, and to the surroundings by dark radiation. The good thermal conductivity of the supports and of the plates gives rise to a temperature gradient between the centre of the strip and its ends. The central junctions are therefore the hot junctions of the thermopile and the ends of the thermopile are the cold ones. The thermopile is covered by two concentric glass hemispheres having inner diameters of 26 and 46 mm respectively. Their thickness is 2 mm. The instrument is shown in Fig. 7.20.

The amount of radiation Q_2 absorbed by the thermopile is given by

$$Q_2 = Q_1 \alpha q r \tag{7.11}$$

where Q_1 is the amount of radiation incident vertically on the receiver surface, α is the absorption coefficient of the black coating of the thermopile, q is the transmissivity, and r the reflectivity of the glass. When Q_2 is a function of $\gamma_\odot$ and $A_\odot$, this expression must be replaced by

$$Q_3 = Q_1 \alpha q r \cdot f(\gamma_\odot, A_\odot) \tag{7.12}$$

where $f(\gamma_\odot, A_\odot)$ is a function of $\gamma_\odot$ and $A_\odot$. The temperature gradient between the centre of the strip and its ends, Δt, is approximately proportional to Q_3. For a more strict treatment of the thermal processes in this kind of instrument Δt must be regarded not only as a function of Q_3, but also of the temperature of the cold junctions (which is equal to the temperature of the instrument).

Under steady-state conditions

$$Q_3\, b\, \mathrm{d}x + \lambda A\, \frac{\mathrm{d}^2(\Delta t)}{\mathrm{d}x^2}\, \mathrm{d}x - K\, \Delta t\, C\, \mathrm{d}x = 0 \qquad (7.13)$$

where b is the width of a strip, $\mathrm{d}x$ is the element of length along the strip, λ is the mean thermal conductivity of the strip material, A is the cross section of the strip, K is the transfer coefficient representing heat losses by conduction, convection and radiation per unit area per unit time per degree, and C is the circumference of the strip. In the above expression, $Q_3 b\, \mathrm{d}x$ is the amount of energy absorbed by the surface element $b\, \mathrm{d}x$ per unit time, $A(\mathrm{d}^2\, \Delta t/\mathrm{d}x^2)\mathrm{d}x$ is the amount of heat passing through the cross-sectional

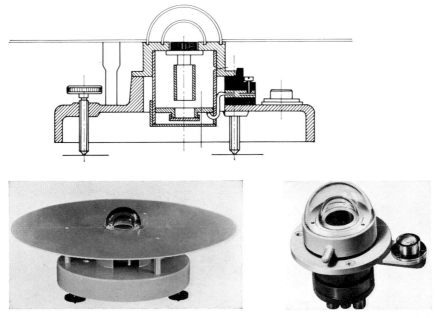

Fig. 7.20. Moll–Gorczynski pyranometer. Top: cross-sectional drawing. Bottom left: mounted. Bottom right: unmounted.

References pp. 312–316

area A of the strip, and $K \Delta t\, C\, dx$ is the heat loss by the element in all possible ways. The boundary conditions must be prescribed before Eq. (7.13) can be solved. A detailed mathematical treatment is available in the literature [54].

Examination of the properties of this instrument yields the following results:

(1) Unevenness in the surface of the thermopile is found to give rise to a departure from the exact cosine response. Deviations of up to 2% are not unusual. Since the thermopile consists of long narrow thin strips, the non-uniformity of one or more of the strips could give rise to a dependence on the azimuth $A_\odot$, especially when the incident radiation changes from parallel to the length of the strip to perpendicular. Tests have shown the presence of a maximum deviation of 2.4% from the mean calibration factor when $\gamma_\odot$ varies from $12°$ to $60°$ and $A_\odot$ from $77°$ to $275°$. Fig. 7.21 shows the results of another examination of a pyranometer of this type, mounted on a rotatable frame [55].

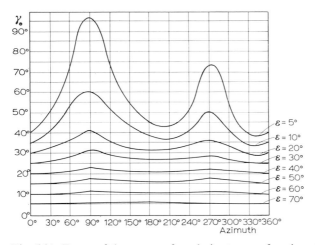

Fig. 7.21. Errors of the output of a solarimeter as a function of $\gamma_\odot$ and azimuth $A_\odot$, for constant values of the irradiation intensity and various values of the angle of incidence ε. Output for $A_\odot = 0$ taken as 100 [55].

The variation seems to be due to changes in the mutual orientation of the parallel beam of light and of the strips, and by the caustic of the glass cover. Starting with $\gamma_\odot = 80°$, the caustic point falls on the thermopile, giving rise to additional local heating. When the incident radiation is parallel to the strips, this point falls on the cold junctions, and when it is perpendicular to the strip, it reaches the hot junctions. This gives rise to a diurnal variation in the calibration factor which is illustrated in Fig. 7.22.

Since the global radiation consists of the direct solar radiation I and the scattered radiation D, the two cosine responses must be considered separately.

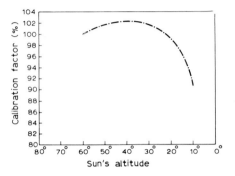

Fig. 7.22. Diurnal variation in the calibration factor.

(2) The temperature coefficient of this instrument depends on a number of factors. The first of these is the temperature dependence of K which is given by

$$K = K_0(1 + K_t t + K_{\Delta t} \Delta t) \qquad (7.14)$$

where K_0 is the heat transfer coefficient at $0°$ and $\Delta t \to 0$, K_t is the temperature coefficient of K, and $K_{\Delta t}$ is the temperature-difference coefficient of K. In addition, the thermal emf and the conductivity of the strips are functions of temperature.

Experimental examinations of the overall temperature coefficient of the calibration factor, in so far as they have been carried out with sufficient care, show an increase of 0.18–0.24% per $1°C$. This temperature coefficient is mainly due to the low conductivity of the air under the glass cover. When the air is replaced by hydrogen, the conductivity is higher than the conductivity of air by a factor of 8. The temperature coefficient of the calibration factor is then almost negligible [54].

(3) Changes in the calibration factor with time are mainly due to deterioration (peeling) of the black and white coatings on the receiver surface but no numerical results are available at present.

(4) The absorption and reflection coefficients of the glass covers are functions of the wavelength. This dependence for $q \cdot r$ is shown in Fig. 7.23. The steep rise in the product $q \cdot r$ at wavelengths between 2900 and 3500 Å is deceptive because the energy reaching the earth in this spectral region is only a small part of the total radiation. A different calibration factor must be used for G-radiation and D-radiation because their spectral compositions are different. The percentage of energy in the above spectral region is greater for D than for I. The calibration factor is lower for the D-component. For a cloudless or a totally overcast sky, the difference in $q \cdot r$ for D and I is about 0.3 %.

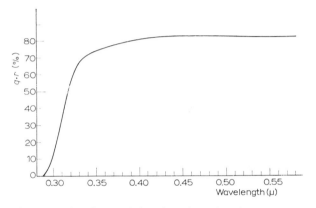

Fig. 7.23. Plot of $q \cdot r$ as a function of wavelength.

(5) Both the volume of the strips and their thermal capacity are small, and the maximum output is reached in less than seconds.

(6) The zero point should be the reading of the instrument when no solar radiation falls on the receiver. When the receiver, the glass cover and all the other parts of the instrument are at a temperature which is equal to the temperature of the cold junctions and to the temperature of the surroundings, the reading of the instrument will represent the 'true zero point'. This condition will not in general be fulfilled. In particular, when the outer glass cover or the metal parts of the instrument lose heat in any way, a 'zero-point depression' will be present. The new zero point is referred to as the 'working zero point'. This applies to measurements of D-radiation, when the glass covers are not heated by the radiation (because they are shaded). The outer glass hemisphere cools down by exchange of radiation with the

surroundings, while the inner hemisphere, which is initially at a higher temperature, radiates to the outer hemisphere. The dark radiation of the inner parts of the instrument, which is emitted in the direction of the aperture is absorbed by the inner glass which is heated as a result, and so on. The cumulative effect of all these processes is that the temperature of the inner glass sphere becomes lower than that of the cold junctions. The heat balance equation (Eq. (7.13)) must therefore be modified to include a term representing the heat loss responsible for the zero-point depression. The heat flow from the thermopile is directed partly upwards, *i.e.* towards the inner glass cover, and partly downwards, towards the supports and the plates. It may be assumed that the temperature of the cold junctions is equal to that of the supports. The temperature difference between the thermopile and the inner glass cover, Δt_{rg}, gives rise to additional cooling of the thermopile. When all these effects are included, Eq. (7.13) becomes

$$Q_3 \, b \, \mathrm{d}x + \lambda A \, \frac{\mathrm{d}^2(\Delta t)}{\mathrm{d}x^2} \, \mathrm{d}x - K \, \Delta t \, C \, \mathrm{d}x - \frac{h}{2} \Delta t_{rg} C \, \mathrm{d}x = 0 \qquad (7.15)$$

where h is the heat transfer coefficient for both directions (upward and downward). The numerical value of the zero-point depression depends on the cloudiness and on the intensity of the G- and the D-radiation. It may reach 25 % of the D-intensity on very clear days.

It is evident from the above discussion that the working zero point must be determined accurately if a correct evaluation of the readings is required. Zero-point depression is of great importance in the calibration of the instruments; for purposes of calibration G- and D-radiation are measured alternately. In the case of G-radiation the receiver is freely exposed to the radiation, but in the case of the D-radiation it is shaded from direct radiation. The working zero point for the D-measurements depends on the time between the instant of shading and the completion of the measurement of D. If the D-reading is taken after a period of shading equal to that necessary for attaining the maximum output for G, it is found that the error caused in the calibration factor is approximately 0.2 %.

(7) Insufficient data are available at present on the dependence of the calibration factor on the amount of incident radiation.

(8) External influences are important. Thus, wind gives rise to considerable cooling of the outer glass hemisphere, which affects the temperature of the inner parts of the instrument as explained above.

The Eppley pyranometer

In this instrument the receiver consists of two concentric rings as shown in Fig. 7.24. The outer ring is white and the middle ring is black; the inner circle is not active. The rings are made of silver foil, 0.25 mm thick, and the coatings are Parson's black and magnesium oxide respectively. A thermal insulator is inserted between the rings. The dull black surface absorbs almost all the radiation incident upon it, while the magnesium oxide reflects visible and near-infrared radiation. As a result, a temperature difference is produced between the rings. Both rings are good absorbers for very long-wave radiation, which is an advantage because long-wave radiation from the glass bulb does not affect the output.

Two models of this pyranometer are available. One of them incorporates 10 junctions and the other 50 junctions. The thermocouples consist of 0.04 mm diameter wires (90% Pt + 10% Rh and 60% Au + 40% Pt). The junctions in thermal contact with the black ring are the hot junctions while those in contact with the white ring are the cold junctions. The temperature difference between the rings gives rise to a thermal emf of the order of 2 mV per cal·cm^{-2}·min^{-1} in the 10 junction model, and 7–8 mV in the 50 junction model. The internal resistance of the two models is 35 and 100 ohms respectively. The upper surface of the receiver lies in the plane passing through the centre of the glass bulb, which is 7.6 cm in diameter and contains dry air.

Let us examine this instrument in the light of the eight criteria listed above:

(1) The calibration factor does not normally depend on the azimuth in view of the circular form of the receiver and its axial symmetry. When this

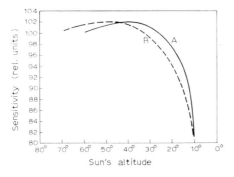

Fig. 7.25. Variation of the sensitivity of two Eppley pyranometers [56]. A: Eppley No. 1571. B: Eppley No. 1220.

dependence is found to be present, it is due to the receiver being incorrectly levelled or not flat. Another source of error is the non-uniformity of the glass which may act as a lens and focus the radiation onto the receiver surface. When $A_\odot$ varies in the range $0°$–$360°$ and $\gamma_\odot$ is $25°$ or more, the variation in the calibration factor is not more than 1–2%. The effect of changes in $\gamma_\odot$ on the calibration factor is a composite one. Possible contributions are: changes in the reflectivity of the receiver coatings with the angle of incidence, changes in the receiver temperature with the amount of incident energy, non-uniformity of the glass, and variation in the position of the caustic point of the receiver.

Fig. 7.25 shows an example of measurements carried out with two Eppley pyranometers by the National Bureau of Standards in Washington and by the Service du Rayonnement, Uccle (Belgium) [56].

The following Table [145] gives experimental results for the cosine response of an Eppley pyranometer calibrated under laboratory conditions with the aid of a lamp giving an intensity of $0.85 \text{ cal} \cdot \text{cm}^{-2} \cdot \text{min}^{-1}$ when incident normally on the receiver surface:

$\gamma_\odot$ (degrees)	90	80	70	60	50	40	30	20	15
Mean deviation (percent)	0	0	−1	−2	−3	−3	−4	−6	−8

The results with an instrument calibrated in sunlight (mean $\gamma_\odot = 50°$) are as follows:

$\gamma_\odot$ (degrees)	90	80	70	60	50	40	30	20	15
Mean deviation (percent)	+3	+3	+2	+1	0	0	−1	−3	−5

In practice, the deviations will be somewhat smaller for $\gamma_\odot$ less than $30°$ because of the increase in D-radiation.

From measurements made in conditions of cloudless sky, in South Africa and at Newport, R.I. (U.S.A.), the percentage *errors* given in the above tables might be amended to:

$\gamma_\odot$ (degrees)	90	80	70	60	50	40	30	20	15
calibr. 90°	0	0	−1	−2	−2	−3	−3	−4	−5
calibr. 50°	+2	+2	+1	0	0	−1	−1	−2	−3

This amendment has been completely ignored by others.

(2a) The output emf of the thermocouples is given (in microvolts) by the following expression

$$U = 32.975 \, (t_h - t_c) + 0.038881 \, (t_h^2 - t_c^2) \tag{7.16}$$

where t_h and t_c are the temperatures of the hot and cold junctions respectively. When $t_h - t_c = \Delta t = $ constant, this equation becomes

$$U = 32.975 \, \Delta t + 0.038881 \, \Delta t \, (2t_h - \Delta t) \tag{7.17}$$

$$\left(\frac{\delta U}{\delta t_h}\right) dt_h = 0.07762 \, \Delta t \, dt_h \tag{7.18}$$

This shows that the output increases by $0.07762 \, \mu V$ per thermocouple per °C.

(2b) The effect of the ambient temperature on the calibration factor is such that the calibration factor changes by 0.05–0.1% per °C for ambient temperatures between −50°C and 40°C.

(2c) Experiments have shown that the calibration factor increases with the temperature while the output diminishes. This appears to be a consequence of the diminution in the temperature difference between the two rings [57–59].

(3) No appreciable ageing effect was observed either in the glass bulb or in the coatings of the rings after 30 months of operation.

(4) Experiments have also shown the absence of a spectral selectivity. The glass transmits practically all radiation between 3500 and 20,000 Å.

(5) The response is such that 98% of the maximum output is reached in 20 and 30 seconds in the case of the 10 and 50 junction models respectively.

(6) and (7) The zero point was found to depend on the amount of incident radiation. An increase in the intensity of radiation from 0 to 1.5 $cal \cdot cm^{-2} \cdot min^{-1}$ was found to give rise to a zero-point depression of 0.15 $cal \cdot cm^{-2} \cdot min^{-1}$. No change in the zero point with the ambient temperature has been found. However, alternate measurements of G-radiation and D-radiation with the same instrument may cause a zero-point depression of a few percent. When used with a potentiometer, the zero points for both D and G must be identical.

(8) Since the instrument is completely sealed, external influences are almost negligible. The Eppley pyranometer is often used for radiation measurements on differently inclined planes, in which case the calibration factor is found to depend on the position of the receiver. Thus, a change from the

vertical to the horizontal position may cause a 2.5 to 3.0% decrease in the calibration factor. This is due to the air convection within the bulb.

A number of improvements of this instrument have been recently introduced by the Eppley Laboratories. The actual receiver is similar to that used with the Eppley pyrheliometer which was described on p. 243. The glass cover is demountable. A clear WG-7 glass, which is transparent to radiation between 2850 and 2800 Å, is employed for total radiation. The Schott filters GG-14, OG-1, RG-2 and RG-8 are used for yellow, orange, red and dark-red radiation respectively. The instrument is mounted on an azimuth elevation stand.

The 'Star' pyranometer (solarimeter)

This widely used instrument was developed at the Zentralanstalt für Meteorologie und Geodynamik (Vienna) [60]. One of the versions of the instrument is illustrated in Fig. 7.26.

The receiver consists of 32 small copper plates which are 0.05 mm thick and are alternately painted black and white. They are attached to two con-

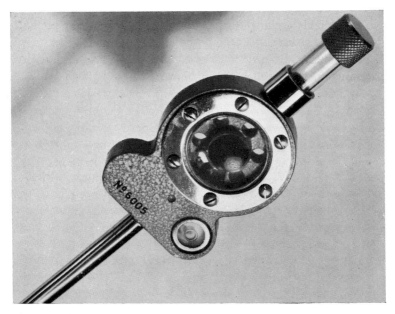

Fig. 7.26. The 'Star' pyranometer.

centric rings which are 3 mm thick and are made of a poorly conducting material. The rings are mounted on a heat-insulating plate, from which they are spaced a certain distance. The thermopile consists of manganin–constantan or copper–constantan junctions which are soldered to the plates. Junctions in thermal contact with black plates are the hot junctions while those in contact with white plates are the cold junctions. The receiver is covered by a polished glass hemisphere which is 2–3 mm thick and 70 mm in diameter. In the spectral region under investigation, the transmissivity of the glass is independent of wavelength, and about 90% of all the incident radiation is transmitted by it.

The incident radiation gives rise to a temperature difference between black and white plates and this generates an emf of about 1.8 mV per $cal \cdot cm^{-2} \cdot min^{-1}$. A larger model of this instrument incorporates 36 copper plates mounted on a nichrome–constantan thermopile, the resistance of which is 3 to 5 ohms.

The properties of the instrument in the light of the aforementioned eight criteria are as follows:

(1a) There is no dependence on $A_\odot$ because of the symmetry of the receiver with respect to its centre, provided the surface is flat and correctly levelled.

(1b) As far as the effect of $\gamma_\odot$ is concerned, there is no departure from the cosine law, down to $\gamma_\odot = 15°$. With a special anti-glare treatment of the white paint, this can be extended to even smaller values of $\gamma_\odot$.

(1c) Finally, it was found that the caustic reaches a distance of approximately $R/2$, where R is the radius of the glass cover [45]. If the dimensions of the glass cover and the receiver are suitably chosen, the measurements are independent of the effect of the caustic.

(2) The overall temperature coefficient of the calibration factor represents changes in the emf of the thermopile, in its conductivity, and in the conductivity of the copper plates, all of which are practically within the limits of accuracy of the instrument.

(3) Ageing was found to occur only during the first six months and appeared as a 2–7% reduction in the sensitivity. No further ageing was observed after this period. The sensitivity was found to remain constant for many years when the receiver was well protected from moisture.

(4) The dependence of the output on wavelength which is usually due to the white coating and the cement added to it, was eliminated by an appropriate choice of materials.

(5) The time response is such that in the small model 99.5 % of the maximum output is attained in less than 20 seconds.

(6) The zero-point drift is very small.

(7) The output is accurately proportional to the amount of radiation incident on the device.

(8) The effect of extraneous influences is small, *e.g.* a change in the wind velocity of 15 m/sec over a period of 5 minutes produces no detectable effect. The calibration factor is independent of position and hence the pyranometer may be used in any position.

Information on other types of pyranometers is available in the literature [61,62].

7. INSTRUMENTS FOR THE MEASUREMENT OF SKY RADIATION

Measurements of the D-radiation are performed on the horizontal plane and hence the same instrument can be employed for this purpose as in the case of the G-radiation, provided a suitable device is used to prevent direct solar radiation from reaching the receiver. The following factors affect the accuracy of such measurements:

(1) A part of the circumsolar sky radiation, which is the most intense part of the D-radiation, is prevented by the shading device from reaching the receiver, to an extent depending on the type of the shading device.

(2) The dimensions of the receivers may not be adequately standardized so that the percentage of circumsolar radiation which does reach the receiver is different for different instruments even though they may incorporate the same shading device.

(3) Multiple reflection within the glass covers affects the accuracy of the measurements.

Other difficulties are encountered in the computations, such as the non-uniformity of the circumsolar radiation and of the D-values throughout the sky. Correct adaptation of all the dimensions of the receiver, and standardization, are therefore far from being a simple task. The problem is aggravated by the fact that the same type of shading is used for different receivers, at different distances.

In addition, stations engaged in measuring G-, I-, and D-radiations, and using them for calibration purposes, may introduce a further element of error by using different opening angles for pyrheliometers and pyrano-meters. As D is only a part of G, the accuracy of the D measurements will be higher than that of the G measurements.

Two types of shading devices are in use, namely, movable and fixed. The movable type consists of a small disc driven by an equatorial mounting and casting a shadow on the receiver. The fixed type consists of a ring or frame which constantly shades the receiver. The advantage of the disc lies in its well-defined shadow which is only slightly affected by the non-uni-formity of the sky. However, it shades off no more of the circumsolar and sky radiation than is essential. In order to utilize its advantages, a very precise equatorial drive is necessary. The disadvantage of this type of shading device is the constant need for supervision, since the smallest departure from the correct motion of the disc gives rise to considerable errors in the measure-ment of the D-radiation. The disc is usually (though not always) made so that it shades the glass cover completely in order to avoid internal reflection phenomena which vitiate the results. The advantage of the ring, on the other hand, lies in the stability, which means that constant supervision is unneces-sary. When the ring is employed, corrections must be introduced for the part of the sky radiation which is obstructed by the sides of the ring. The com-putation is not simple because of the non-uniformity of the sky even for a cloudless day, and is further complicated by the presence of partial cloudiness.

Fig. 7.27. Shading ring as used by the U.S. Weather Bureau [63].

The ring arrangement used by the U.S. Weather Bureau is shown in Fig. 7.27. The ring has a radius of 20 inches and a cross-sectional diameter of 2 inches. The inclination of the frame is equal to the geographical latitude at the particular point, and the shading is modified in accordance with the date by a parallel displacement of the ring. This involves a change in the distance between the ring and the receiver, and hence different parts of the circum-solar radiation are included in the results. This is more noticeable when the cross-sectional diameter of the ring is smaller than the diameter of the glass cover, *e.g.* in the case of the ring used with the Eppley pyranometer.

It is clear that the internal reflection will also be affected. This may be avoided by means of a shading device ensuring a fixed shading–receiver distance throughout the year [64–67]. This device is shown in Fig. 7.29. The shading strips form a part of a sphere so that the distance from the central line of each strip to the centre of the receiver remains constant throughout the year and a shadow of constant width and shape is cast on the instrument.

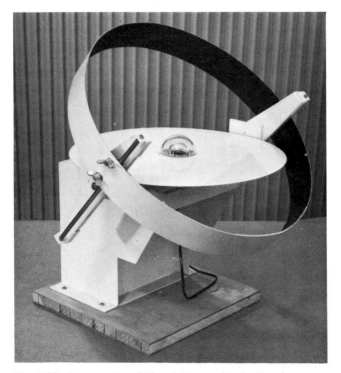

Fig. 7.28. Pyranometer of Kipp & Zonen with shading ring.

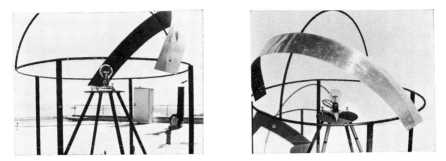

Fig. 7.29. Shading ring ensuring fixed ring-receiver distance.

The dimensions of the shading device depend on the aperture of the pyrheliometer used at the particular station for measuring I-radiation. Since, as was noted above, the shading ring intercepts not only the direct radiation but also a proportion of lateral sky radiation, a correction for this must be introduced in the final analysis of the results obtained for the D-radiation. The correction can either be computed or determined experimentally. An approximate formula for the correction is available in the literature [68–70]. Experimentally, the correction can be obtained by successive measurements with the ring and a disc of suitable diameter. The amount of the D-radiation which is cut off is added to the results obtained with the ring alone. For totally overcast sky, measurements with and without the ring yield the necessary correction.

So far, instruments for the measurement of I-, G- and D-radiation have been discussed: a universal instrument has also been described in the literature [71].

8. CALIBRATION OF SOLAR RADIATION INSTRUMENTS

When systematic measurements of I-, G- and D-radiation are available, the sun may be used as the 'source' and calibration may be performed with the aid of the formula

$$I = \frac{\bar{G} - \bar{D}}{\sin \gamma_\odot} \tag{7.19}$$

where the bars indicate averages over the period of calibration. The intensity of I-radiation obtained in this way can be checked from time to time against a sub-standard.

The use of this formula is not simple, since many precautions and corrections are required. Thus cloudiness is a very important factor, as Eq. (7.19) applies to a totally clear sky.

For accurate calibration, all the criteria mentioned in the foregoing paragraphs for pyrheliometers and pyranometers must be carefully examined. In addition, the opening angles of the pyrheliometer and the pyranometer must be the same when shaded. Unless all the measurements are carried out with extreme care, the errors may accumulate to the extent of making the calibration worthless [69].

9. DISTILLATION INSTRUMENTS

In various branches of agriculture, botany, zoology, human biology and so on, the radiation by the surroundings is of importance in addition to the direct solar and sky radiation. The total radiation plus the radiation reflected from the surroundings, as intercepted by a spherical receiver, is called the circumglobal radiation.

The Bellani spherical pyranometer

The first instrument designed for the determination of circumglobal radiation was invented by Bellani in 1836 and was modified by Henry in 1926. This instrument operates on the principle of distillation of a liquid as a result of the absorption of the incident radiation.

The Bellani instrument was first considered theoretically by Borrel [72] who tried to compute its sensitivity as a function of the height of alcohol in the tube (Fig. 7.30), the temperature, and the absorptivity of the alcohol inside the transparent globe. Wierzejewski and Prohaska [73,74] developed a theory for an instrument with a dark globe.

The Swiss model. An important improvement of the Bellani pyranometer was introduced by the Physikalisch-Meteorologisches Observatorium, Davos–Platz, Switzerland [75]. This modification consists of covering the inner glass globe with a grey metal coating. In Fig. 7.30, A is alcohol, B is an evacuated opaque glass globe, C is the second glass globe, D is a glass tube with its open end E above the maximum possible height of the alcohol in the globe and F is an extension of this tube which is calibrated in cubic centi-metres. The absorbed incident circumglobal radiation heats the alcohol and gives rise to a partial evaporation of the latter. Some of the evaporated alcohol condenses mainly above the tube because the vapour pressure of the alcohol in the tube is lower than that in the globe. The amount of condensed alcohol in tube F is a measure of the amount of radiant energy entering the inner globe.

It is evident from Fig. 7.30 that the container and the alcohol have a considerable thermal capacity. This means that a certain amount of time must elapse before steady-state conditions are reached. This is particularly noticeable at the beginning of the operation. For example, with a constant intensity of radiation, 90% of the final indicator reading is attained after 45 minutes following the beginning of the exposure. This response time is reduced when the amount of alcohol in the globe becomes smaller. Measure-

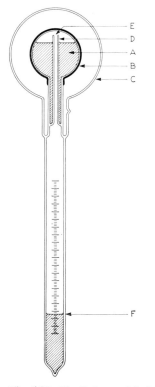

Fig. 7.30. The Swiss model of the Bellani pyranometer.

ments carried out over short periods of time will clearly be inaccurate. A suitable period is at least half a day. Whole-day exposures yield satisfactory results when recorded after sunset when the heat accumulated in the instrument has all been used up for evaporation. The instrument is simple to handle and is cheap. The calibration factor is found to depend on the temperature and is usually given in the maker's calibration certificate. Attempts have been reported in the literature to find a conversion factor from G-radiation to circumglobal measurements [76–79].

The British model designed by W. E. Wilson is a further modification of the original Bellani pyranometer. Its operation is based on the transfer of heat absorbed by a black body to an evaporating liquid. Again, the amount of condensed liquid is a measure of the incoming radiant energy. Other improvements have been reported elsewhere [80–84]. A recent model is shown in Fig. 7.31, in which the sphere is made of copper. It is coated with lamp black and is evacuated. There are various versions of this device. Standardization appears to be impossible.

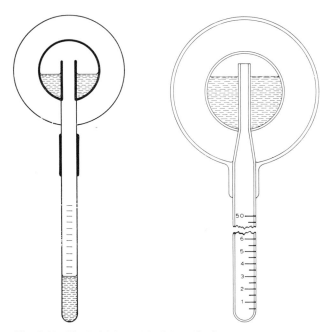

Fig. 7.31. The British model of the Bellani pyranometer.

Fig. 7.32. The German model of the Bellani pyranometer.

In order to obtain a more uniform evaporation it was suggested to wet the inner surface of the glass container. This idea was incorporated in *the German model* (Fig. 7.32) in which the inner wall of the bulb is coated with glass powder. The bulb itself is made of black glass and is highly evacuated [85]. The calibration factor increases with the amount of distilled alcohol. The glass-powder lining of the black globe improves the constancy of the calibration factor. Results obtained with instruments of similar construction and containing alcohol of the same degree of purity have been found to give good agreement.

Distillation instruments with flat receivers

When high accuracy is not demanded, it would be advantageous to replace expensive pyranometers for the measurement of G-radiation by instruments which would be as simple in operation and as cheap as the Bellani pyranometer. In such an instrument the receiver would have to be flat and horizontal. The need for a cheap instrument has been emphasised by UNESCO and other international institutions during the sessions of the International Radiation Commission. This would allow an increase in the number of solar stations all over the world.

Two models of this kind will be mentioned here. The first is being developed at Davos and the second at the Observatoire Météorologique, Trappes, France. The latter model is known as Type Météorologie Nationale S.282.B. The radiation is absorbed by the blackened flat surface and the heat flow through the aluminium envelope covering a glass container. Both the container and the aluminium cover are protected from heat losses by a plastic foam jacket. The flat receiver is covered by two concentric hemispheres.The distillation and condensation proceed as in the other distillation instruments. When the temperature is constant during the time of exposure, the radiant energy is used for distillation only. If, however, the temperatures of the container and the calibrated tube at the beginning and the end of the exposure are different, then additional distillation takes place and vitiates the results. In order to exclude this source of error, another similarly constructed instrument, having a flat white receiver, is employed. The difference in the readings of the two instruments is proportional to the absorbed amount of energy. Should the air temperature increase more rapidly than that of the container, the distillation may take place in the opposite direction, *i.e.* from the calibrated tube into the container.

10. MEASUREMENT OF THE DURATION OF SUNSHINE

Many solar stations are provided with sunshine duration recorders only. The importance of such recorders lies not only in providing the actual duration of sunshine and the actual times of sunshine, but also in the possibility of using the data provided by these recorders for estimating the amount of insolation and forecasting of insolation.

The Campbell–Stokes sunshine recorder

This instrument was invented in England by Campbell in 1853 and was modified by Stokes in 1879. It is essentially a polished solid glass sphere with an axis mounted parallel to that of the earth (Fig. 7.33). The sphere acts as a lens and the focused image moves along a specially prepared paper bearing a time scale. Bright sunshine burns a path along this paper.

Fig. 7.34a shows a path recorded in this way for a day with bright sunshine. The result for a day with broken cloud is shown in Fig. 7.34b. The precise evaluation of these records is not at all simple. If the path is of the kind shown in Fig. 7.34c (burnt through), then the length L corresponding to the duration of sunshine is given by $L = l - (d_1 + d_2)/2$. When the sunshine is

Fig. 7.33. The Campbell–Stokes sunshine recorder.

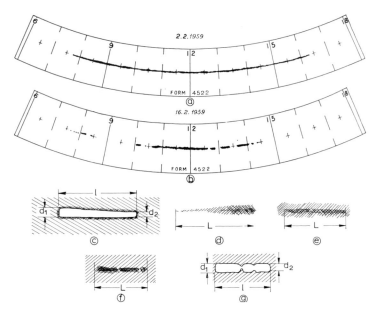

Fig. 7.34. Typical recordings obtained with the Campbell–Stokes sunshine recorder.

not strong enough to burn the paper right through there will be a darker core with a light brown track (Fig. 7.34d). In this case the total length L is taken. This also applies to the case shown in Fig. 7.34e.

A very short period of sunshine gives rise to a small spot with or without a darker core. When one or two spots are found, the time of sunshine is assigned a symbol 0, meaning less than 0.05 of an hour of sunshine. Three or four spots are given the symbol 0.1 of an hour, five or six spots 0.2 of an hour, and so on.

Other kinds of paths show what are known as 'constrictions', seen best on the back of the paper (Fig. 7.34f). If one or two well developed constrictions, or three poorly developed ones, are formed during one hour, 0.1 hour is subtracted. Similarly, 0.2 hour is subtracted if three well developed constrictions are present. The time subtracted in this way must not exceed 0.4 of an hour. When numerous constrictions appear, the distances between them are measured and added up.

A burned hole may be interrupted by a white time-mark on the paper when the sunshine is too weak to overcome the high reflectivity of these marks. Such interruptions are disregarded.

The sunshine duration tables should be scaled in 0.1 hour intervals. A little zero (0) denotes no sunshine being recorded at all, whilst a capital zero (O) stands for less than 5 % sunshine.

The theoretically possible duration of sunshine N and the measured duration n should be noted for every day, as well as $(n/N) \cdot 100$. Monthly totals should also be computed for the same hourly intervals.

If the sunlight is intercepted by obstacles, the latter must be determined for every station. The time of interception must then be included in n [146].

Additional difficulties may be caused by inaccuracy of the charts (which differ from supplier to supplier, and sometimes even from batch to batch (Fig. 7.34g)), and by improper installation of the instrument. Three possible sources of error should be mentioned [86]:

(a) Azimuth deviation: deviation of the axis from the north–south direction.

(b) Pole deviation: the axis does not point exactly at the North Pole.

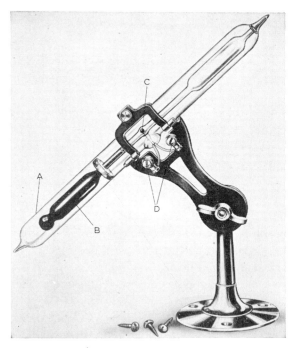

Fig. 7.35. The Martin–Marvin sunshine recorder.

This may be due to an incorrect setting of the instrument's axis to the geographical latitude, or to uneven installation in the north–south direction after the latitude has been set correctly.

(c) Inaccurate levelling in the east–west direction, resulting in unevenness.

The Martin–Marvin thermoelectric sunshine recorder

This instrument is shown in Fig. 7.35. In this device A is a glass tube protecting the instrument, B is a glass container with mercury and C are two contacts connected to the terminals D. The duration of sunshine is determined as the time for which the contacts are shorted by the mercury [87,88]. The advantage of this instrument is that the recording instrument may be used inside a building or in a remote place. The Martin–Marvin instrument is less sensitive than the Campbell–Stokes recorder [89]. Both are insensitive to low sunshine and cannot be used to record times of sunrise and sunset.

Photoelectric sunshine recorder

A photoelectric recorder of increased sensitivity has been developed by the U.S. Weather Bureau [90]. It is shown in Fig. 7.36 and incorporates two photoelectric cells mounted in a hermetically sealed glass cylinder. One of these cells is shielded by the outer ring which is adjusted in accordance with the data, while the other is exposed to the sun. In the absence of direct sunshine, the common reading is zero. An electric current flows as soon as solar radiation falls on the device. This current trips the sensitive relay K_1 and this operates either a recorder or a totalising device.

A comparison between this recorder and the Martin–Marvin recorder shows that on clear days the former instrument records approximately one hour more of sunshine. The photoelectric recorder begins to operate approximately 40 minutes earlier than the Martin–Marvin and is capable of indicating the presence of broken cloud, which is not registered by the other recorder. On days with a completely overcast sky the photoelectric recorder records even two hours more of sunshine.

Sunshine duration and insolation

Ångström [91] was the first to find a correlation between the theore-

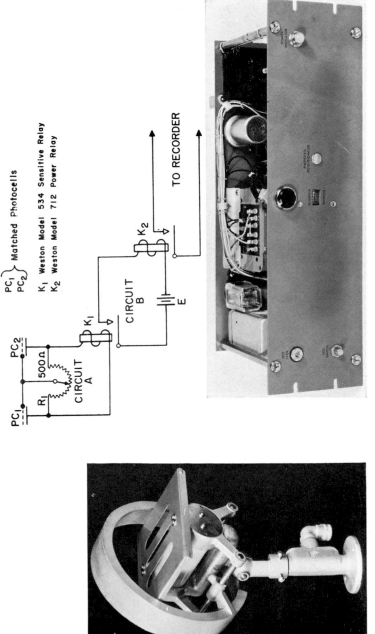

Fig. 7.36. Photoelectric sunshine recorder. Left: external view. Top right: circuit diagram. Bottom right: recording unit.

tically possible hours of sunshine N, the actual duration of sunshine n, the actual amount of global radiation G, and the global radiation on a day with a clear sky G_m. This relation is

$$G = G_m[a' + (1-a')n/N] \tag{7.20}$$

where a' is the mean proportion of the global radiation for a completely overcast sky.

Another expression has been reported by Fritz and MacDonald [92] for monthly average values of G and n/N. This expression is

$$G = G_m(0.35 + 0.61\, n/N) \tag{7.21}$$

Daily values are given by

$$G/G_m = K + (1-K)\, n/N \tag{7.22}$$

where K is the ratio of total radiation without sunshine to that with 100% sunshine. The quantity n/N is not a constant and a factor which is a function of n/N must be introduced into Eq. (7.22) to allow for this. The final result is

$$G/G_m = K + (1-K)Cn/N \tag{7.23}$$

Empirical values of Cn/N are constant for given n/N at all latitudes whereas K is very dependent on both latitude and season. Yet another expression which has been employed is

$$G/G_0 = a + b\, n/N \tag{7.24}$$

where G_0 is the value of G for a completely transparent atmosphere. In this expression b is virtually constant but a is a function of latitude. The empirical formula for the latter dependence is

$$G/G_0 = 0.29 \cos \varphi + 0.52\, n/N \tag{7.25}$$

It holds for latitudes between 0 and $60°$. The sum of actual duration of sunshine and the time of cloudiness must be equal to 100% but this is not always the case.

Other correlations between the duration of sunshine and the values of G have also been reported [93–106].

11. MEASUREMENT OF SOLAR RADIATION WITHIN LIMITED SPECTRAL REGIONS

Three methods are practicable for measuring radiation within limited spectral regions. They are: (a) the use of spectrographs and monochromators, (b) the use of receivers which are sensitive within broader spectral limits than required, or even of spectrally unselective receivers, in combination with suitable filters, and (c) the use of instruments having receivers sensitive only to the spectral region under investigation.

Spectrographs and monochromators

The use of spectrographs and monochromators is limited to specialized laboratories and will not be discussed in detail. Fig. 7.37 shows an arrangement used in the author's laboratory, for the continuous recording of energy distribution in the I-spectrum. Recently the instrument was improved: instead of a photomultiplier solar cells and a recorder are being used. The recorder activates the drum D in such a way that through the exit slit S_2 a narrow part of the spectrum passes step by step till the entire spectrum has passed through it.

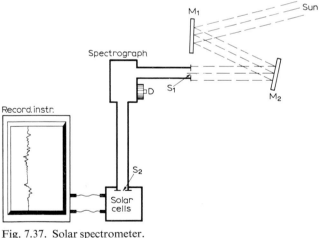

Fig. 7.37. Solar spectrometer.
M_1 rotating mirror of the coelestat
M_2 fixed mirror of the coelestat
S_1 entrance slit of the spectrograph
S_2 exit slit of the spectrograph
D drum to rotate the prism of the spectrograph

Filters

Optical filters are often used in order to separate out the required spectral region. In addition to the so-called grey or neutral filters, which have the same relative transmittance at all wavelengths, there are also the so-called cut-off and interference filters. They transfer different percentages of energy within different wavelength regions.

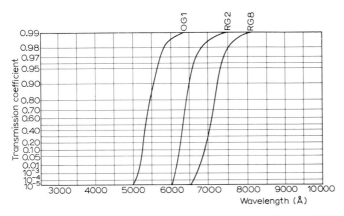

Fig. 7.38. Characteristics of the Schott–Jena OG–1, RG–2 and RG–8 filters.

Cut-off filters (see Fig. 7.38) are made of coloured glass and owe their properties either to a mixture of coloured ions in the glass mass, or to the presence of sub-microscopic crystals. In the first case the spectral transmittance depends on the kind of ions present and in the second on the composition and dimensions of the crystals. Another type consists of dyes in gelatine, the film being either lacquered on or cemented between flat glass plates. Liquid filters have also been used. These filters reflect and absorb a part of the radiation, the reminder being transmitted (see [107–109]).

The reflectance r_λ at normal incidence and given wavelength is given by

$$r_\lambda = \left(\frac{n_\lambda - 1}{n_\lambda + 1}\right)^2 \quad \text{or in } \% : \quad r_\lambda = \left(\frac{n_\lambda - 1}{n_\lambda + 1}\right)^2 \cdot 100 \qquad (7.26)$$

where n_λ is the refractive index of the filter material at the particular wavelength λ. Another quantity which is often employed is the coefficient of reflection R_λ which is given by

$$R_\lambda = 1 - 2\,r_\lambda \qquad (7.27)$$

The factor 2 represents the fact that the radiation is reflected from both surfaces of the filter. The ratio of the transmitted energy to the incoming energy is called the transmittance τ_λ of the filter. The ratio of the transmittance τ_λ to the coefficient of reflection R_λ is known as the coefficient of transmission θ_λ. The ability of a filter to transmit a certain fraction of the incoming radiation is its optical density which is given by

$$\delta = \log_{10} \tau_\lambda \tag{7.28}$$

Both τ_λ and θ_λ for a glass cut-off filter depend on five factors, namely: thickness, angle of incidence, temperature, age, and the time of exposure. Let us consider these effects in turn.

The coefficient of transmission is related to the thickness by the expression $d = const \log \theta_\lambda$. This can also be described in terms of an extinction factor which is defined by $\varepsilon_\lambda = \log \theta_\lambda$ so that $d = const \, \varepsilon_\lambda$.

The dependence on the angle of incidence which can be described by a factor of the form $(1 - \sin^2 \alpha / n^2)^{-\frac{1}{2}}$ is due to the increased path length within the filter with increasing angle of incidence. The main effect of a rise in temperature is a shift in the shortwave limit towards longer wavelengths. Finally, the transmission maximum decreases with age and the time of exposure to the sun.

In the case of liquid cut-off filters the concentration of the solution changes with temperature, and with the spectral composition of the radiation absorbed by the solution and the exposure time (see below). According to Beer's law, the transmission coefficient of liquid filters is given by $\theta = 10^{-cd/d_1}$ where c is the concentration, d is the thickness of the layer, and d_1 is the thickness of a layer of unit concentration and 10% transmittance. Changes in the temperature may give rise to dissociation of the soluble components of the solution, while the absorbed radiation may take part in photochemical processes within the solution. As the time of exposure increases, the by-products of the photochemical processes are found to accumulate.

In order to obtain a desired pass-band, a combination of cut-off filters may be employed. For any filter combination, the final value of the transmission coefficient is equal to the product of the transmission coefficients of the separate filters. The overall pass-band may be determined either experimentally or by calculation.

Interference filters are based on the principle of the Fabry–Perot interferometer. Details of this type of interferometer may be found in standard textbooks on physical optics.

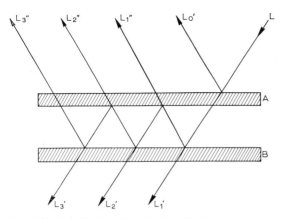

Fig. 7.39. Principle of interference filter.

A cross-section of such a filter is shown in Fig. 7.39. The glass plates A and B are coated with thin semi-transparent layers of silver and are separated by a distance equal to an odd number of half-wavelengths. The incident beam is partly reflected at the outer surface of A and partly transmitted by it. The radiation transmitted by A is multiply reflected between the plates with the result that some of it leaves the system through the outer face of A, while the remainder passes through the outer surface of B.

Beams L_1', L_2', ... and L_0'', L_1'', ... will therefore be out of phase, and will interfere, giving rise to spectra of different orders. Beams with other phases are not transmitted. A narrow band of the spectrum may be selected by suppressing all other regions with glass or additional interference filters.

An ideal interference filter for a given spectral region $\Delta\lambda$ wavelengths wide ($\pm\frac{1}{2}$ $\Delta\lambda$ from the centre of this region) would be one having a rectangular transmittance curve with $\Delta\lambda$ as the base; in other words, it would transmit equally all wavelengths within the region of and none outside this region.

The quality of an interference filter is characterized by its maximum transmittance (τ_{max}) of λ_m, the centre of the transmitted band, by the shape of its transmittance curve, and by the absence of other spectral regions.

The shape of the transmittance curve is usually defined by the so-called half-width ($H.W.$), expressed by the wavelength range measured from the centre of the curve ($\tau_\lambda = \tau_{max}$) to the point at which $\tau_\lambda = \tau_{max}/2$ (Fig. 7.40).

Two additional points used for more detailed characterization of the transmittance curve are $\tau_\lambda = \tau_{max}/10$ and $\tau_\lambda = \tau_{max}/100$.

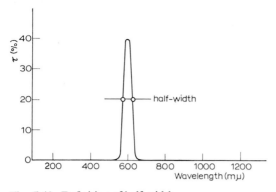

Fig. 7.40. Definition of half-width.

The overall curve may be determined for a given interference filter from known values of λ_m, τ_{max} and $H.W.$ If, for example, $\tau_{max} = 30\%$, $H.W. = 100$ Å, and $\lambda_m = 5900$ Å, the transmittance within the filter's acceptance band, say at 6150 Å, can be calculated. In this case $d\lambda = 250$ Å; $d\lambda/H.W. = 250/100 = 2.5$. From data tabulated for the filter, we then obtain for $d\lambda/H.W. = 2.5$, the value $\tau_{6150\ \text{Å}} = 0.0035 \times 30\% \sim 0.1\%$ [111].

The wavelength corresponding to the maximum of the transmission curve depends on the angle of incidence of the radiation. With increasing angle of incidence, the wavelengths corresponding to the maximum of the transmission curve shifts towards higher wavelengths, and the pass-band gradually develops into two lines, both of which are polarized. It is, therefore, possible to change the wavelength corresponding to the maximum of the transmission curve by merely altering the angle of incidence.

If a narrow well-defined spectral region is required, care must be taken to ensure that divergence of the incident beam is small.

For spectral regions which are broader than those dealt with so far,

Fig. 7.41. Multilayer interference filter.

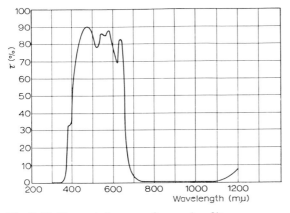

Fig. 7.42. Transmission curve for an edge-filter.

another kind of interference filter known as the *multilayer interference filter* is employed. This is illustrated in Fig. 7.41. Layers marked I have a higher refractive index than those marked II [112]. The required interference is achieved by a suitable choice of the thickness of each layer and the total number of layers. The interference is produced after multiple reflection at the surfaces of the various layers.

The filter whose transmission curve is shown in Fig. 7.42 is known as an *edge filter*.

A particular advantage of interference filters, as compared with cut-off filters, is that in the former there is very little absorption by the filter so that its temperature does not rise. This is not the case for cut-off filters whose temperature does tend to rise.

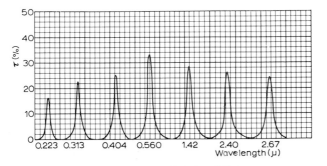

Fig. 7.43. Characteristic of a multiple band filter.

References pp. 312–316

Another very useful type of filter which can also be used as a kind of spectrograph is a *multiple band filter*, or the so-called *wedge interference filter*. The transmission of this type of filter is shown in Fig. 7.43 [113].

The use of filters in solar radiation measurements

Let us now consider the uses of the various filters described above. The filter characteristics which will be discussed below relate to filters alone and not to the receivers with which they are used.

Cut-off filters for measurement of I-radiation are mostly used in order to divide the I-radiation energy into spectral regions. Table 7.2 gives some data for three filters which have been adopted and recommended internationally (see Fig. 7.38, p. 283).

TABLE 7.2. Characteristics of three Schott–Jena filters.

Filter	Thickness (mm)	Spectral region (Å)
OG–1	2.4	5250–28,000
RG–2	1.5	6300–28,000
RG–8	3.0	7100–27,000

Despite very careful selection by the Davos Observatory, some departures from the standard are found. The Davos Observatory carries out tests on filters and supplies in each case a reduction factor (D.R.) which is used to bring the final results to within 0.5% of the transmittance of a standard filter. Small differences in the thickness of individual filters are taken into account in this reduction factor which is supplied with the filter. Large differences in thickness (1–4 mm) have recently been investigated by Ångström and Drummond [114]. They found the following relation

$$\lambda_\mu = \lambda'_\mu + \frac{\ln l}{\gamma} \tag{7.29}$$

where λ_μ is the wavelength at 50% transmittance, λ'_μ is its value for a thickness of 1 mm, l is the thickness of the filter and $\gamma = 0.14, 0.13$ and 0.11 for OG–1, RG–2 and RG–8 respectively. With increasing temperature, λ_μ shifts towards longer wavelengths. For a $10°C$ temperature rise this shift is 1.2, 1.4 and $1.8\ m\mu$ for the OG–1, RG–2 and RG–8 filters respectively. These shifts are independent of the thickness within the range 1–5 mm. A constant diminution in the maximum transmittance by OG–1 and RG–2 filters has been found to amount to respectively 12.4 % and 12.6 % for a 5-year period [115].

The reduction factor must sometimes be adapted to the kind of pyrheliometer with which they are used. This is because the three filters can be employed with pyrheliometers having (1) no cover over the aperture, (2) a glass cover, or (3) a quartz cover, and the D.R. values are strictly accurate only for instruments of type (1).

If a pyrheliometer which is normally covered by a glass or quartz window to protect the receiver surface from dust and moisture is used uncovered for calibration, the resulting I-values must be corrected for the additional reflection by multiplying the D.R. by 1.08. Conversely, in the case of a pyrheliometer covered at all times with a glass or quartz window, the D.R. value must be divided by 1.08.

The I-measurements may be evaluated in a number of spectral regions. Let I_t be the total energy measured by the pyrheliometer having a quartz window, or none at all, I_1, I_2, and I_3 the energies measured with OG–1, RG–2, and RG–8 respectively, and $DR1$, $DR2$, and $DR3$ the corresponding reduction factors. The energy in the various spectral regions can be calculated; some results are shown in Table 7.3.

TABLE 7.3. The energy in various spectral regions, evaluated from measurements with three different filters [116].

Limits of λ (Å)	Evaluation
$\lambda < 5250$	$I_t - DR1 \cdot I_1$
$5250 < \lambda < 6300$	$DR1 \cdot I_1 - DR2 \cdot I_2$
$\lambda < 6300$	$I_t - DR2 \cdot I_2$
$5250 < \lambda < 7100$	$DR1 \cdot I_1 - DR3 \cdot I_3$
$\lambda < 7100$	$I_t - DR3 \cdot I_3$
$\lambda > 6300$	$DR2 \cdot I_2$
$\lambda > 7100$	$DR3 \cdot I_3$

The filters described so far were all flat and unsuitable for G- and D-measurements. Hemispherical filters of the same optical transmittance as the three filters discussed above have been used by Marchgraber and Drummond in the Eppley pyranometer (cf. Fig. 7.24, p. 262). It was found possible to separate the G-radiation into spectral regions with the aid of sensitive recording instruments and filters, just as in the case of the I-radiation.

Spectrally selective receivers

Phototubes depend on the external photoelectric effect which consists of the emission of electrons from a suitable surface as a result of absorption of radiation. The photoelectric effect is described by Einstein's equation

$$h v = hc/\lambda = \tfrac{1}{2} m v^2 + W \tag{7.30}$$

where h is Planck's constant, v is the frequency of the incident radiation, c is the velocity of light, m is the mass of an electron, v is the velocity of the photoelectrons and W is the work which must be done on the photoelectrons to take them out of the solid. In order that the photoelectric effect should take place, it is clearly necessary that the energy of the incident radiation should be greater than W. It follows that the above equation may be rewritten in the form

$$h(v - v_0) = \tfrac{1}{2} m v^2 \tag{7.31}$$

where v_0 is the minimum frequency for which the photoelectric effect will occur. The number of photoelectrons emitted at a given frequency is a function of the intensity of the incident radiation and of the frequency.

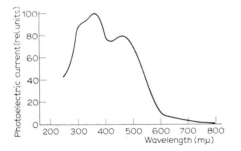

Fig. 7.44. Spectral characteristic of an Sb–Cs phototube.

In addition to the normal photoelectric effect described above, some metals, and especially alkali metals, exhibit a selective photoelectric effect. In this case the photoelectric current is found to increase with increasing frequency until a maximum is reached. Further increase in the frequency gives rise to a rapid reduction in the photoelectric current. There may be one or more secondary maxima (Fig. 7.44).

In practice, a phototube consists of a bulb of ordinary or special glass or of silica, depending on the spectral region in which the tube is to be used

(Fig. 7.45). The photocathode C is in the form of a metal coating deposited on the inner surface of the bulb, and the anode A is usually in the form of a rod, ring or grid. The bulb is evacuated or filled with an inert gas. The electrons emitted by the photocathode are accelerated by the battery B. The resulting electron current is measured by a galvanometer G.

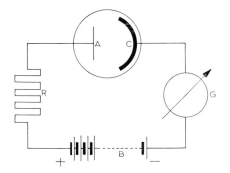

Fig. 7.45. Circuit for a phototube.

The voltage–current curve for a vacuum phototube is shown in Fig. 7.46 (curve a). So long as all the emitted electrons are not collected, the current increases with increasing voltage. Eventually all the electrons are collected and the saturation current is recorded.

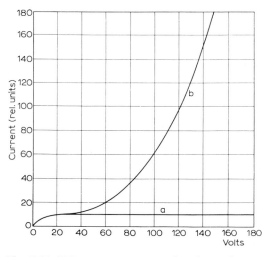

Fig. 7.46. Voltage–current curves for phototubes.
(a) Vacuum tube. (b) Gas-filled tube.

References pp. 312–316

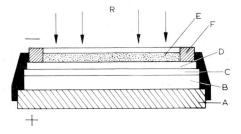

Fig. 7.47. Construction of a selenium photocell.

In the gas-filled tube, the moving electrons give rise to ionization and excitation of the gas. This leads to an increase in the output current and there is no saturation (Fig. 7.46, curve b).

Vacuum phototubes are less sensitive than gas-filled tubes although they do have a more stable output and spectral sensitivity. At a given wavelength and applied voltage, the output current of a vacuum tube is very nearly proportional to the incident intensity. This is not so in the case of gas-filled tubes.

Photocells of the barrier-layer type make use of a semi-conducting material, *e.g.* selenium, and are described in Chapter 8.

Fig. 7.47 shows a cross-section of a selenium photocell. In this figure A is a metal plate which serves as a base for the selenium layer B (a p-type semiconductor), C is an n-type semiconductor, D is a conducting and transparent layer, E is a layer of lacquer, F is a metal ring and R denotes the incident radiation. The plate A, whose thickness is of the order of 1 mm, has a

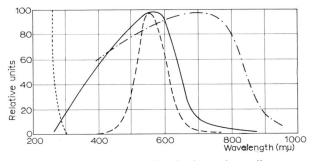

Fig. 7.48. Spectral response of barrier-layer photocells.
. . . . . . Cadmium photocell
——— Selenium photocell
—·—·—· Infrared cell
- - - - - Human eye

double function. It supports the selenium layer and also serves as the positive electrode of the photocell. D is transparent to the radiation and has a good electrical conductivity. It is in contact with F which serves as the negative electrode. The action of this type of cell cannot be described in terms of Ohm's law. The characteristics of the barrier-layer cells depend on some 10 variables and will now be considered in detail.

The spectral response of barrier-layer photocells is shown in Fig. 7.48. In a limited temperature range, the dependence of the emf of a barrier-layer photocell on the temperature may be described by

$$U = U_0(1 + \alpha t) \tag{7.32}$$

where the temperature coefficient α depends on the energy, its absolute value decreasing with increasing energy as shown in Fig. 7.49.

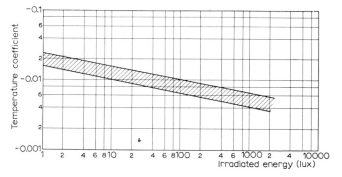

Fig. 7.49. Temperature coefficient for a barrier-layer photocell as a function of irradiated energy.

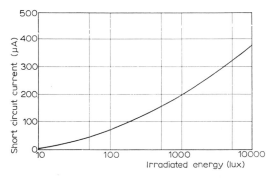

Fig. 7.50. Short-circuit current of a barrier-layer photocell as a function of irradiated energy.

References pp. 312–316

Fig. 7.50 shows the short-circuit current as a function of irradiated energy in lux. The short-circuit current is also a function of temperature. As a first approximation, the temperature dependence may be described by a linear formula of the form

$$i_{s,t} = i_{s,0} \left[1 + \frac{\alpha R_e t}{R_i + R_e} \right] \qquad (7.33)$$

where R_e and R_i are the external and internal resistances. When $R_e \ll R_i$, the temperature coefficient β decreases with R_e and is temperature independent.

Regular barrier-layer cells with flat sensitive surfaces follow the cosine law up to angles of incidence of about 45°. The time lag is exceptionally short and may in fact be neglected. Only in the case of alternating light beams of high frequency does the electric capacitance of the cell (10^5 pF · cm^{-2}) cause any difficulty. Ageing and fatigue are caused by prolonged exposure to strong radiation, but, under proper working conditions, the fatigue effect may be neglected. Cesium and germanium are used in addition to selenium, in view of their sensitivity in the infrared range (germanium is sensitive from 7000 to 19,000 Å).

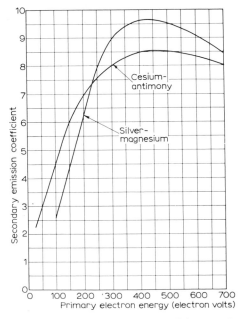

Fig. 7.51. Secondary emission coefficient as a function of primary electron energy.

Electrons emitted as a result of any particular process may produce further electrons, known as secondary electrons, when they strike a metal or a dielectric, provided their energy is sufficiently high. The secondary emission coefficient ϱ is defined as the ratio of the primary to secondary electron currents. Fig. 7.51 shows this coefficient as a function of the primary electron energy in electron-volts; ϱ increases rapidly with increasing primary electron energy, but if the latter is sufficiently great to penetrate deeply into the bombarded material, ϱ decreases because the secondary electrons are prevented from escaping.

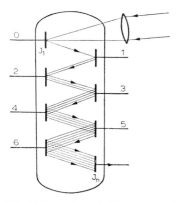

Fig. 7.52. Schematic illustration of a photomultiplier.

Photomultipliers are devices in which a number of targets are arranged so that secondary electrons in one target are used to bombard the next. This type of device is a phototube in which the electron current increases with each target, giving rise to self amplification (Fig. 7.52).

Ultraviolet measurements

Measurements in the regions 2800–3800 Å, 3150–3800 Å, and 2800–3150 Å can only be carried out by means of complicated and expensive spectrometers using specially trained personnel. These instruments are not suitable for field work. An example of this type of apparatus is shown in Fig. 7.53.

In the spectra of I- and D-radiation reaching the earth, the energy decreases towards the ultraviolet. The receiver having a reverse characteristic, *i.e.* the receiver whose sensitivity increases with decreasing wavelength, is desirable for ultraviolet measurements.

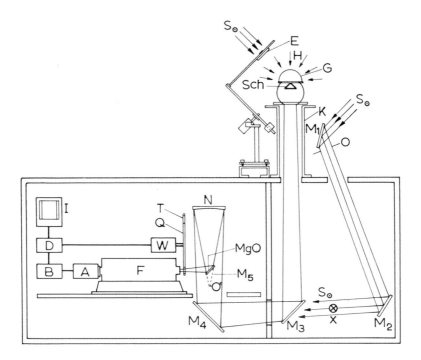

Fig. 7.53. A recording spectrograph [149].

A   Photomultiplier box (tube E.M.I. type 6256)
B   D.C. amplifier
D   Automatic sensitivity adaption
E   Screen for shading off direct solar radiation
F   Double monochromator
G   Quartz envelope of diffuser sphere
H   Incoming sky radiation
I   Recorder
K   Steel tube
M_{1-5} Plane mirrors
MgO MgO whitening cover of mirror M_5 for recording direct solar radiation
N   Spherical mirror
O   Diaphragm for solar radiation
O′   Optical image of O
Q   Disc for neutral filters
$S_\odot$ Direct solar radiation
Sch   Diffuser sphere with internal cone for obtaining correspondence with the cosine law (developed by Larche and Schulze)
T   Neutral filters mounted on Q for the adaption of the sensitivity
W   Mechanism turning the disc Q
X   Position of standard lamp when the apparatus is calibrated for direct solar radiation

Any decision regarding the instrumentation for ultraviolet measurements must therefore be made taking account of the degree of accuracy required.

The energy reaching the earth in the ultraviolet range varies considerably as a function of γ, turbidity, cloudiness, the seasons, and with the elevation above sea level.

The effects of ultraviolet radiation on higher and lower forms of life are very individual. Thus if applicability and effects of the ultraviolet are the aim, rather than a precise study of the radiation itself, the instrument need not be of very high grade, and an accuracy of 10% or even worse may be quite acceptable. In such cases, phototubes, photocells, and photomultipliers can be used, if they are at least checked against a standard ultraviolet lamp or compared frequently with a similar ultraviolet meter.

Cadmium phototubes as used for ultraviolet measurements, have rather individual characteristics. They differ from each other and must be checked individually. In addition, ageing gives rise to changes in the specific output, spectral sensitivity, and the voltage–current characteristics, leading to the formation of electric charges on the walls of the tubes. One of the sources of these disturbing factors is gas adsorption on cadmium surfaces. It follows that thorough degassing is necessary in order to achieve adequate stability. Higher stability may be realized with cadmium cathodes sensitized with water vapour.

In view of these difficulties, cadmium tubes are more suitable for calibration than for use in the field. The drawbacks are then of lesser importance and their effect on the final results may be reduced by the so-called filter quotient method. In this method two measurements are carried out with the same phototube. In one of them the tube is used without the filter, giving the total intensity, and in the other the filter is incorporated. The filter quotient Q, which is equal to the ratio of the two values obtained in this way is then evaluated. Experience shows that the calibration factor of a cadmium phototube may be determined as a function of Q to an accuracy of 2 to 5% by comparison with a standard tube. The relation frequently employed is

$$f(Q) = c \cdot Q^b \text{ or } \log f(Q) = \log c + b \log Q \qquad (7.34)$$

where c and b are constants.

Cadmium phototubes have been used and investigated in detail at the Davos Observatory [117–120].

A spectral sensitivity curve for cadmium is shown in Fig. 7.54.

Titanium phototubes have been mainly used by the National Bureau of Standards in Washington. Their spectral characteristic is shown in Fig. 7.55. The output is usually calibrated in terms of absolute units [121,122].

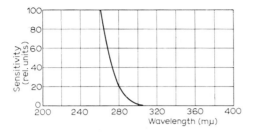

Fig. 7.54. Spectral sensitivity curve for cadmium.

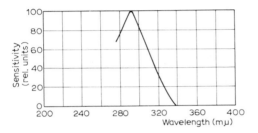

Fig. 7.55. Spectral sensitivity curve for titanium.

An ultraviolet meter used at the Israel Institute of Technology is shown in Fig. 7.56.

Other ultraviolet instruments have also been described in the literature [123–125]. Receivers which can be used without filters for the separate determination of radiation in the ranges 3150–3800 Å and 2800–3150 Å are not available.

Ultraviolet meters with selenium photocells. An ultraviolet meter for the 3150–3800 Å range, incorporating a selenium photocell, has been used since 1951 in conjunction with filters for the range 3000–4000 Å by Schulze in Hamburg (see Fig. 7.57). The filters employed are UG–5-10 mm and WG–5-2 mm (Schott u.Gen.). In order to preserve the cosine response, the whole receiver is covered by a Larché sphere. Schulze has designed an instrument with a sodium photocell and a special filter for the range 2800–3200 Å (Bäckström filter plus picric acid plus UG–5-10 mm). Both instruments can be used for continuous recording [126–128].

Fig. 7.56. Ultraviolet meter.

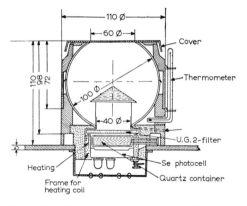

Fig. 7.57. Ultraviolet meter incorporating a selenium photocell (after Schulze).

For climatological purposes an ultraviolet-sensitive photovoltaic cell (GEC type PV–1) can be used in conjunction with a Corning glass filter (7–37). The PV–1 cell is sensitive up to 2500 Å with a maximum transmission (32%) at 3600 Å. The transmittance of the filter is zero at 3150 and 3800 Å.

Measurements in the visible range

The radiation which can be perceived by the human eye lies, roughly speaking, between 3800 and 7800 Å. Fig. 7.48 (p. 292) shows spectral sensitivity curves for photocells and for the human eye. Adaptation of sensitivity to the human eye can, if necessary, be achieved by means of special filters (the Weston Company manufactures the 'Viscor' filter which can be used for this purpose).

Meters using photocells, with an output calibrated in foot–candles or lux units, can be used for taking either momentary or continuous readings of the illumination intensity; the instruments are known as photometers or foot–candle- or lux-meters. The term 'illumination' used for light corresponds to 'insolation' or 'irradiation' used for other spectral regions.

The criteria of a reliable photometer are constancy of the output with time (lack of ageing or fatigue), high reproducibility, good cosine response, accuracy in weak and strong light, and an absence of a temperature-dependence.

Studies carried out on selenium photocells show that their sensitivity falls off as a result of ageing and fatigue, the latter being due to exposure to very strong illumination (whilst 1000 lux is the highest intensity which does not lead to fatigue [129], the illumination may in some places be as high as 150,000 lux).

Neutral filters must therefore be used to protect the photocells. The filters supplied by different manufacturers differ from each other, and it is advisable to check the photometer's output frequently for different illumination intensities, at a constant temperature, by means of a standard lamp. The latter should preferably be covered with a filter converting the light to 'white light'.

The cosine response is improved by a slightly convex milk-glass filter covering the surface of the cell. This gives a good cosine proportionality down to 3°.

Some photometers designed by the Radiation Commission will now be briefly described, in alphabetical order.

Electrocells. These photometers have a selenium photocell, which can be used for either momentary or prolonged measurements of the illumination intensity. Both are calibrated in lux units. To avoid temperature and moisture effects, the photocells may be supplied sealed in a thermostat. The scale is calibrated logarithmically from 100 to 14,000 lux, the range being from 25 to 14,000 lux. The cells are artificially aged, and are protected from strong illumination with a special layer. The spectral sensitivity is adapted to that of a human eye by means of a filter. The photocell is covered with a milk-glass plate, and is protected by a polished glass hemisphere. The complete instrument shows no temperature-dependence [130]. No information is available regarding the previously mentioned criteria.

Megatrons. In this case the selenium photocell is protected with a spherical opal perspex lens (cover) for cosine correction, and with a neutral filter to reduce the intensity of the incident light, so that linearity of the output is preserved [131] even for strong illumination. An electrolytic integrator can be supplied with these instruments to adapt them for continuous measurements. The cosine response is correct for all angles of incidence [132].

Weston instruments. The Weston model 756 with cosine and spectral correction is suitable for the measurement of outdoor light intensities between 0 and 12,000 foot–candles, in ranges of 0–120, 0–1200, and 0–12,000 foot–candles [133]. It is advisable to have two calibrated cells with the same indicator, one cell for everyday use and the other for checking. The latter should be stored in a dry dark place, and used as a standard for the former; it may be checked against a standard lamp. For measuring sky brightness, the photocell should be mounted on rotatable stands for $A_\odot$ and $\gamma_\odot$ adjustment.

Infrared measurements

Infrared measurements may be carried out with the aid of spectrally unselective receivers and filters. Since this part of the spectrum contains

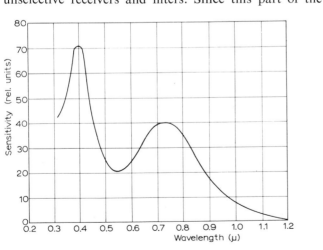

Fig. 7.58. Spectral sensitivity curve for cesium.

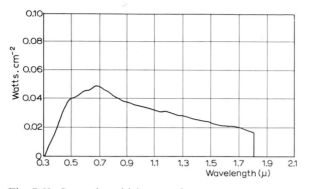

Fig. 7.59. Spectral sensitivity curve for germanium.

about 30% of the total energy, the difficulties encountered in the case of ultraviolet measurements do not arise. Cesium and germanium photocells are suitable as spectrally selective receivers for the infrared (see Figs. 7.58 and 7.59).

12. ALBEDOMETERS

The albedo A is defined as the ratio of the amount of radiation R reflected by an object to the energy falling on the same object. Instruments which are used for the determination of G-radiation can also be used as albedometers provided they are equipped with a device for turning them up and down. The two main sources of error for uniform surroundings are: the different spectral composition of the reflected and incident radiation, and the difference in the angles which are subtended during the measurement of the reflected and incident radiation. In order to avoid complicated corrections, the instruments should incorporate a receiver and a cover which are as spectrally unselective as possible. While the reflected radiation is very nearly the diffusely reflected radiation, the incident radiation is partly directional and partly diffuse, so that the angle of irradiation is of some importance.

The value of the reflected radiation depends also on the shape of the reflecting surface, which interferes with a precise numerical determination of the albedo at certain points.

If the surroundings are of a different character from the surface whose albedo is to be measured, the effect of these surroundings on the value of A must be taken into consideration. For the calculation of a case of symmetrical surroundings, see [134].

13. LONG-WAVE RADIOMETERS AND BALANCE METERS

The radiation components I, G, D and A discussed till now belong to the 'short-wave' radiation (cf. Table 1.1, page 1). When however radiation exchange between short-wave radiation, the atmosphere, and the ground is considered, radiation other than short-wave must also be taken into account.

All radiation components exhibit two characteristics: direction (downwards or upwards) and spectral range (long-wave or short-wave). These components are:

↓Downward: I, G (or D)–short-wave solar radiation

   ↓L–long-wave atmospheric (return) radiation

↑Upward: R_I, R_G (or R_D)–reflected parts of I, G (or D), short-wave

   ↑L–terrestrial dark radiation, long-wave

   r–reflected part of ↓L

Radiation in which all possible components participate is known as a 'complete radiation balance':

$$Q_1 = S + D + {\downarrow}L - E - R_s - r \tag{7.35}$$

where S is the sun + sky radiation on any plane other than that normal to the beam, D is the diffuse radiation on this plane, E is the terrestrial emission and R_s is the upward-reflected part of S.

When the sun is covered:

$$Q_2 = D + {\downarrow}L - E - R_D - r \tag{7.36}$$

At night, the long-wave components alone remain:

$$Q_3 = {\downarrow}L - E - r \tag{7.37}$$

This balance is known as 'long-wave' or 'infrared' balance.

Instruments for the various kinds of balances must be designed in such a way as to be capable of reacting to all radiation components composing the balance under consideration.

Instruments having a horizontal receiver for the various Q-values will now be described and discussed.

Many such instruments were developed and some models of them are in use; an attempt was made to coordinate them, when, as a result of discussions of the Radiation Commission of the IAM (IUGG) and the Working Group of WMO, and following an invitation from the Meteorological Service of the Federal Republic of Germany, comparisons between the most commonly used instruments were instituted in 1955 and 1956. Detailed information on these comparisons is laid down in two mimeographed reports of the Hamburg Meteorological Observatory (1955 and 1956); short abstracts are published in the WMO Bulletin [147, 148].

These instruments can be classified either according to the components which they measure, or by the principles of their construction and operation. The classification according to components is as follows:

(1) 'Radiation balance meters'–for all radiation components.

(2) 'Infrared balance meters'–for the long-wave components alone.

(3) 'Effective pyranometers'–for the short-wave and long-wave components incident on and reflected from a black surface directed upward or downward.

(4) 'Infrared effective pyranometers'–as (3), but for the long-wave components only.

As these instruments are pure radiation meters, all heat exchange other than that due to radiation (such as convection) must be excluded. This may be accomplished by setting up an artificial wind, which provides constant forced convection greater than the expected natural convection. Another method is to cover the receiver with a material transparent to all radiation components which are to be measured. The cover must be in thermal equilibrium with the receiver when measurements are made. In some instruments, a third method is applied: two surfaces (receivers), one of them shaded from radiation, are equally exposed to convection, the shaded surface being heated to compensate for the heat-exchange of the irradiated surface.

Radiometers can thus be divided into three groups: (1) uncovered, ventilated; (2) covered with transparent covers; (3) uncovered, with compensation.

Only the one instrument which serves as a standard, and some others, which are commercially available, will be discussed; other types will merely be mentioned. Some common properties will be examined with regard to the following points:

(1) The components to be measured.

(2) Spectral properties of the black dye and of the cover in unventilated instruments.

(3) Cosine response.

(4) The kind of measurements: momentary, continuous recording with or without integration.

(5) Units of output and auxiliary instruments.

(6) Accuracy.

(7) Inertia.

(8) Dependence on the ventilation.

(9) Dependence on the wind velocity.

(10) Dependence on the air temperature.

(11) Influence of rain, dew, and hoar frost.

(12) Calibration procedure.

Although the nomenclature used by the constructors for manufacturers is not always ideal, it will be adopted in the following discussion.

Balance meter, Davos model (after P. Courvoisier)

Fig. 7.60 shows a general view of this instrument [135–137]. The receivers are horizontal, blackened circular plates, one of which is directed upwards and the other downwards. Ventilation is used to stabilize the heat flow. The air flow is laminar. A special feature of the ventilation arrangement is a third stream of air flowing between the two glass plates, excluding any heat transfer between them.

The twelve criteria may be summed up as follows:

(1) The instrument can be used as a balance meter, as an effective

Fig. 7.60. Four-component radiation balance meter, Davos model 1962.

pyranometer, or a device for the determination of the soil temperature at night.

(2) No information about the spectral properties of the black dye and of the cover is available.

(3) No information about the cosine response is available.

(4) Its readings represent instantaneous values only.

(5) Its output is in the millivolt range.

(6) The overall accuracy is quite high, the total of the errors not exceeding 1% (this does not include the difference between the short- and long-wave absorption power of the black dye.

(7) The response time is such that maximum output is obtained in 1–1.5 minutes.

(8) An air speed of 4–12 m/sec gives rise to a change of only a few percent in the output.

(9) The influence of wind depends on velocity and direction. If the direction of the wind is opposite to that of the air stream, or if it is lateral, then its velocity must not exceed 3–4 m/sec. In the case of a strong wind, the instrument should be turned in the direction of the wind.

(10) The device does not exhibit any noticeable dependence on the air temperature.

(11) Equal heating of the two plates eliminates effects associated with dew and hoar frost. No protection from the rain is provided.

(12) The sun is used as a source for calibration and the measurements are carried out with the sun unobstructed and obstructed respectively. The intensity of insolation is measured separately.

Net exchange radiometer and total hemispherical radiometer (after J.T. Gier and R. V. Dunkle)

These instruments are illustrated in Fig. 7.61 [138]. The receivers of both instruments are of the same construction and are ventilated by identical blowers, giving rise to equal air streams on both sides of the receivers. The only difference between the two instruments is that in the net exchange meter both the upper and the lower surface of the receiver are blackened (Fig. 7.61b), whilst in the hemispherical radiometer the lower surface is polished and is protected from radiation by an aluminium shield which is polished outside and blackened inside (Fig. 7. 61c).

The receiver is made of 180 turns of constantan wire (0.04 mm in

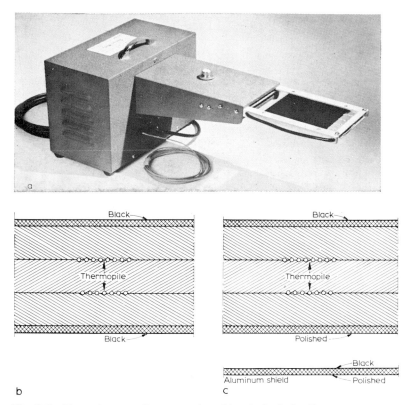

Fig. 7.61. Net exchange radiometer and total hemispherical radiometer.

diameter) which is wound on a bakelite slab (114 × 114 mm). One side of the flat constantan coil is silver plated. In this way a series of 180 thermocouples is obtained. The temperature difference between the hot and cold junctions is proportional to the heat flow through the receiver. The bakelite with the thermopile is covered on both sides with thin pieces of bakelite which in turn are covered by aluminium foil. The thermocouple for measuring the temperature of the receiver is inserted in the centre of the latter.

    The twelve criteria are as follows:

    (1) The net exchange radiometer (balance meter) is used to measure radiation balance. The total hemispherical radiometer is used to measure the effective radiation.

    (2) The black coating employed is the 'Fuller Flat Black Decoret', which possesses different absorptivities in the long- and short-wavelength

regions. The absorptivity is 93% between 1000 and 2000 Å, and 88% between 11,000 and 12,000 Å. In spite of this, the calibration is almost constant for all radiation components.

(3) The cosine response is very good because of the diffuse nature of the surface.

(4) The output can be recorded in dry weather.

(5) The output is usually measured with a millivoltmeter or a recording potentiometer (scale approximately from -5 mV to $+15$ mV).

(6) The accuracy is 5% (this figure is obtained by comparing the net exchange meter with two total hemispherical radiometers, one turned up and the other down, and *vice versa*, with the receivers horizontal.

(7) The inertia is low, *e.g.* 95% of the maximum output is obtained in 12 seconds.

(8) Nothing is known about the influence of ventilation.

(9) The influence of wind depends on the velocity and direction. Thus, a wind of 6.5 m/sec gives rise to an error of about 4%, whilst the same or even stronger wind of up to 13.5 m/sec causes an error of 2% if its direction is the same as the direction of ventilation.

(10) The effect of temperature changes on the accuracy is 0.25% per °C.

(11) Measurements can be carried out in the presence of rain; the formation of dew and hoar frost may be prevented by ventilation.

(12) The calibration is usually carried out against a standard lamp.

The Hamburg–Lupolen instrument (radiation balance meter after R. Schulze)

This instrument consists of two receivers, one facing up and the other down (Fig. 7.62) [139, 140]. Blackened Moll-type thermopiles with an internal resistance of 50 ohms are employed; they consist of circular metallic plates 30 cm in diameter and are covered with white lacquer.

The hot junctions of the thermopile are in thermal contact with the blackened surface and the cold junctions are in contact with the aluminium cylinder which is 60 mm in diameter and 40 mm long. Inside the cylinder there is another thermopile, which is 6 metres long. The hot junctions of this thermopile are in thermal contact with the aluminium cylinder and the cold junctions are located 1 m below ground. The thermopiles at the centre of the metal discs are covered by polythene caps (Lupolen H, 0.1 mm thick), protecting the thermopile from wind, moisture, etc.

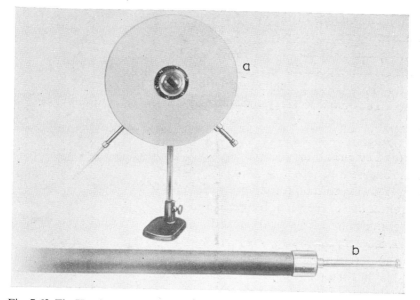

Fig. 7.62. The Hamburg–Lupolen instrument.

The twelve criteria are as follows:

(1) The following components can be measured with the instrument:
a) The upper receiver alone, during sunshine:

$G+D+\downarrow L-\sigma T_u^4$, where T_u is the absolute temperature of the upper thermopile and σT_u^4 the radiant energy emitted by the upper surface; at night: $\downarrow L-\sigma T_u^4$.

b) The lower receiver alone, during sunshine:

$E+R+r-\sigma T_1^4$, where T_1 is the absolute temperature of the lower thermopile and σT_1^4 the radiant energy emitted by the lower surface: at night: $E+r-\sigma T_1^4$.
c) The two receivers connected, during sunshine: Q_1 (see Eq. 7.35); at night: $\downarrow L-E$.

(2) The spectral response of the instrument depends on the properties of the coating of the receivers and of those of the covers. It appears that the black used by Kipp & Zonen is less black for the infrared. The Lupolen is similar to glass for the short-wave radiation, while its transparency to the total infrared radiation is 83%, 6% being lost by reflection and 11% by absorption.

(3) The cosine response is good, owing to the properties of the Lupolen cover.

(4) The outputs of the upper and lower thermopiles separately, the instrument temperature and the output of the thermopiles connected together are recorded on the same chart at intervals of 20 sec. An integrating device can be incorporated.

(5) Output: no information available.

(6) The overall accuracy, including effects associated with the recorder is 3 %.

(7) See (4).

(8) There is no dependence on ventilation because the instrument is covered as described above.

(9) Owing to the protection given by the Lupolen caps, there is no direct effect due to the wind. However, owing to the higher temperature of the near side of the caps, there is an additional dark radiation of about 0.2% which reaches the receiver.

(10) The influence of air temperature on the calibration factor is negligible, but the temperature of the instrument affects the results. This effect amounts to 1.5 % per °C.

(11) Rain has no effect and dew is formed only on the upper cup. Hoar frost is found to interference with the results.

(12) Short-wave calibration is carried out with the sun as the source. For longer wavelengths, the calibration is carried out by placing the instrument in a Dewar flask which can be maintained at various fixed temperatures.

14. INSTRUMENTS FOR THE DETERMINATION OF TURBIDITY AND THE AMOUNT OF PRECIPITABLE WATER IN THE ATMOSPHERE

The procedure used in the determination of turbidity and the amount of precipitable water in the atmosphere consists essentially in measuring the direct solar radiation I with and without filters of suitable spectral transmittance. In the method due to Schüepp, an ordinary normal incidence pyrheliometer and two Schott cut-off filters (RG–2 and OG–1) are employed (cf. Fig. 7.38, p. 283). This method is a development of existing methods and brings this kind of measurement to the level of higher accuracy. Schüepp has also developed a graphical procedure for the determination of α, β, B and w independently of the particular pyrheliometer and filters employed [141] (see p. 150).

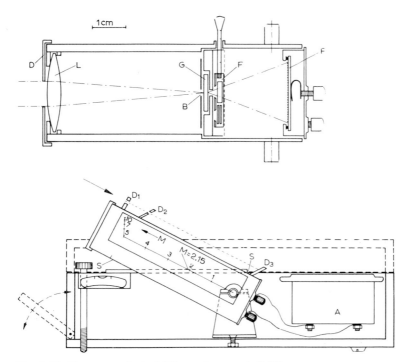

Fig. 7.63. Photometer for turbidity measurements [142].

Fig. 7.63 shows a new type of sun photometer for turbidity measurements (after F. Volz) [142]. The lens L has a focal length of 45 mm. The opening of the diaphragm D gives maximum output of the microamperemeter A during strong sunshine. This diaphragm can be removed during periods with low sunshine intensity. In the focal plane of the lens an aperture B of 1.8 mm is located. The amount of circumsolar radiation is negligible. Behind this aperture a diffusing glass plate G is fixed for scattering the solar radiation on the surface of the photocell P. The diopters D_1 and D_2 are for approximate adjustment, D_3 for exact adjustment. Behind the aperture B there is a filter frame F. The microamperemeter A has a range of 25–100 μA; its internal resistance is ca. 1000 ohms. The air mass is defined by the point of intersection of the inclined line M with the horizontal s—s'. Maximum accuracy is achieved for air masses between 2 and 4. The response time is such that maximum output is attained in 2 seconds.

Other methods have also been described in the literature [141,143] which are based on the observation of the absorption by water vapour in the atmosphere (cf. Chapter 3). In this method the solar radiation is dispersed by a grating or prism and the energies measured in two narrow spectral regions, one of which corresponds to a strong absorption band and the other to a region where there is practically no absorption [144]. The amount of precipitable water vapour can thus be determined if the relation between the absorbing mass and the rate of absorption is known. In spectral regions where the absorption is continuous, an exponential function can be employed. The absorption bands of water vapour are of course not continuous.

REFERENCES

1. K. WEGENER, *Geofis. Pura Appl.*, 18 (1950) 2–4.
2. W. SCHÜEPP, *Geofis. Pura Appl.*, 23 (1952) 3–8.
3. P. COURVOISIER, *Arch. Meteorol. Geophys. Bioklimatol. Ser. B*, 5 (1954) 125–145.
4. K. WEGENER, *Geofis. Pura Appl.*, 43 (1959) 2–8.
5. C. G. ABBOT, L. B. ALDRICH AND A. G. FROILAND, *Smithsonian Inst. Misc. Collections*, 123, No. 5 (1954) 1–4.
6. P. COURVOISIER AND H. WIERZEJEWSKI, *Arch. Meteorol. Geophys. Bioklimatol. Ser. B*, 1 (1948) 45–53.
7. P. COURVOISIER AND H. WIERZEJEWSKI, *Arch. Meteorol. Geophys. Bioklimatol. Ser. B*, 1 (1948) 156–199.
8. *Ann. Astrophys. Obs. Smithsonian Inst.*, 2 (1908) 39–47.
9. *Ann. Astrophys. Obs. Smithsonian Inst.*, 3 (1913) 53.
10. *Ann. Astrophys. Obs. Smithsonian Inst.*, 7 (1954) 99–100.

11. W. M. SHULGIN, *Monthly Weather Rev.*, 55 (1927) 361.
12. *Ann. Astrophys. Obs. Smithsonian Inst.*, 3 (1913) 64.
13. K. FEUSSNER, *Meteorol. Z.*, 52 (1935) 318–326.
14. K. FEUSSNER, *Meteorol. Z.*, 53 (1936) 303–307, 361–374.
15. *Ann. Astrophys. Obs. Smithsonian Inst.*, 2 (1908) 36.
16. C. G. ABBOT, *Smithsonian Inst. Misc. Collections*, 56, No. 19 (1911) 1–10.
17. L. B. ALDRICH, *Smithsonian Inst. Misc. Collections*, 111 (1949) 1–11.
18. *Ann. Astrophys. Obs. Smithsonian Inst.*, 5 (1932) 83–85.
19. K. ÅNGSTRÖM, *Phys. Rev.*, 1 (1893) 365.
20. A. ÅNGSTRÖM, *Tellus*, 10 (1958) 342–354.
21. P. COURVOISIER, *Phys. Meteorol. Obs. Davos, Internal Rept.* No. 206 (1959).
22. E. BACKLIN, *Arkiv Mat. Astron. Fysik*, 19 A, No. 10 (1925) 1–14.
23. J. C. THAMS, *Wetter und Leben*, 1 (1948) 143–148.
24. J. C. THAMS, *Mitt. Schweiz. Meteorol. Zentralanstalt*, 81 (1948) 73–84.
25. R. H. ELDRIDGE, *Quart. J. Roy. Meteorol. Soc.*, 78 (1952) 260–264.
26. P. COURVOISIER, *Medd. Sveriges Meteorol. Hydrol. Inst. Ser. B*, 13 (1957) 3–51.
27. F. LINDHOLM, *Tellus*, 10 (1958) 250–255.
28. *Ann. Astrophys. Obs. Smithsonian Inst.*, 6 (1942) 50–53.
29. E. BÄCKLIN, *Arkiv Mat. Astron. Fysik*, 19 A (1925) 1–8.
30. IGY Instruction Manual, Part VI: Radiation Instruments and Measurements, *Annals of IGY*, V (1957), Pergamon, London.
31. F. LINDHOLM, *Geofis. Pura Appl.*, 41 (1958) 177–182.
32. J. GEORGI, *Tech. Mitt. Instrumentenwesen Deut. Wetterdienstes*, 3 (1957) 15–21.
33. L. BOSSY AND R. PASTIELS, *Inst. Roy. Meteorol. Belg., Mem.*, No. 29 (1948).
34. R. PASTIELS, *Inst. Roy. Meteorol. Belg., Publ. Ser. A*, No. 11 (1959).
35. W. MÖRIKOFER, Meteorologische Strahlungsmessmethoden, in *Handbuch der biologischen Arbeitsmethoden*, Urban & Schwarzenberg, Berlin–Vienna, Abt. II, Band 3 (1939), pp. 4057–4066.
36. SAVINOV, *Tr. Gl. Geofiz. Observ.*, No. 14 (1949).
37. K. YA. KONDRATYEV, *Radiant Solar Energy* (in Russian), Leningrad (1954), pp. 32–40.
38. K. YA. KONDRATYEV, *Radiant Solar Energy* (in Russian), Leningrad (1954), p. 39.
39. W. MÖRIKOFER AND CHR. THAMS, *Meteorol. Z.*, 53 (1936) 22–26.
40. P. COURVOISIER, Theorie des Bimetallaktinograph Robitzsch, *Intern. Radiation Conference Rome, 1954* (mimeographed, not printed).
41. W. MÖRIKOFER AND CHR. THAMS, *Meteorol. Z.*, 53 (1936) 409–415.
42. CHR. THAMS, *Ann. Schweiz. Meteorol. Zentralanstalt*, 80, No. 6 (1943) 4–7.
43. W. MÖRIKOFER AND CHR. THAMS, *Meteorol. Z.*, 54 (1937) 360–371.
44. CHR. THAMS, *Ann. Schweiz. Meteorol. Zentralanstalt*, 81, No. 6 (1944) 19–25.
45. H. WÖRNER, *Z. Meteorol.*, 7 (1953) 289–298.
46. J. C. THAMS, *Geofis. Pura Appl.*, 24 (1953) 115–124.
47. M. J. BLACKWELL, *Meteorol. Res. Paper*, No. 791, Meteorol. Office, London (1953).
48. P. COURVOISIER, *Phys. Meteorol. Obs. Davos, Internal Rept.*, No. 166 (1954).
49. J. GEORGI, *Polarforschung*, 20 (1950) 353–356.
50. J. GEORGI, *Ann. Meteorol.*, 4 (1951) 227–236.
51. J. GEORGI, *Meteorol. Rundschau*, 5 (1952) 181–182.
52. J. GEORGI, *Geofis. Pura Appl.*, 20 (1951) 1–11.
53. J. GEORGI, *Veröffentl. Meteorol. Hydrol. Dienst. Deut. Dem. Republik*, 14 (1954) 1–882.
54. P. BENER, *Arch. Meteorol. Geophys. Bioklimatol. Ser. B*, 2 (1950) 188–249.
55. H. HINZPETER, *Z. Meteorol.*, 6 (1952) 118–121.
56. R. DOGNIAUX AND R. PASTIELS, *Inst. Roy. Meteorol. Belg., Publ. Ser. B*, No. 16 (1955).

57. B. B. WOERZ AND J. F. HAND, *Monthly Weather Rev.*, 69 (1941) 146–151.
58. P. COURVOISIER, *Phys. Meteorol. Obs. Davos, Internal Rept.*, No. 121 (1951).
59. T. H. MACDONALD, *Monthly Weather Rev.*, 79 (1951) 153–159.
60. I. DIRMHIRN, *Arch. Meteorol. Geophys. Bioklimatol. Ser. B*, 9 (1958) 124–148.
61. J. GEORGI AND O. HAASE, *Meteorol. Rundschau*, 4 (1951) 188–192.
62. N. RODSKJER, *Arkiv Geofysik*, 2 (1955) 385–393.
63. J. F. HAND, *U.S. Weather Bureau*, Nov. 1946.
64. N. ROBINSON, *Bull. Am. Meteorol. Soc.*, 36 (1955) 32–34.
65. N. ROBINSON AND L. STOCH, *J. Appl. Meteorol.*, 3 (1964) 179–181.
66. N. ROBINSON, *Bull. Res. Council Israel*, 8 (1954) 435.
67. N. ROBINSON, Shading device for instruments for the measurement of sky radiation, *3rd Intern. Congr. Intern. Soc. Biometeorol.*, *Pau 1963* (in the press).
68. M. J. BLACKWELL, *Meteorol. Res. Paper*, No. 895, Meteorol. Office, London (1954).
69. IGY Instruction Manual, Part VI: Radiation Instruments and Measurements, *Annals of IGY*, V (1957), Pergamon, London, pp. 412, 427–428.
70. A. G. DRUMMOND, *Arch. Meteorol. Geophys. Bioklimatol. Ser. B*, 7 (1956) 413.
71. J. GEORGI, *Meteorol. Rundschau*, 9 (1956) 1–4.
72. F. BORREL, *Météorologie*, T 19 (1943) 147–180.
73. F. PROHASKA AND H. WIERZEJEWSKI, *Ann. Geophys.*, 3 (1947) 184.
74. F. PROHASKA, *Ber. Schweiz. Botan. Ges.*, 57 (1957) 101–114.
75. P. COURVOISIER AND H. WIERZEJEWSKI, *Arch. Meteorol. Geophys. Bioklimatol. Ser. B*, 5 (1954) 413–446.
76. W. SCHÜEPP, *Arch. Meteorol. Geophys. Bioklimatol. Ser. B*, 10 (1960) 311.
77. J. C. THAMS AND H. WIERZEJEWSKI, *Arch. Meteorol. Geophys. Bioklimatol. Ser. B*, 9 (1958) 185–198.
78. H. WÖRNER, *Z. Meteorol.*, 10 (1956) 65–75.
79. H. WÖRNER, *Z. Meteorol.*, 11 (1957) 225–233.
80. P. A. BRIXTON, *Proc. Roy. Soc. (London)*, B 96 (1924) 123–131.
81. P. A. BRIXTON, *J. Hyg.*, 25 (1926) 285–294.
82. P. A. BRIXTON, *Mem. London School Hyg. Trop. Med.*, 1 (1927) 22–36.
83. D. L. GUNN, R. L. KIRK AND J. A. H. WATERHOUSE, *J. Exptl. Biol.*, 22 (1945) 1–7.
84. D. L. GUNN AND D. YEO, *Quart. J. Roy. Meteorol. Soc.*, 77 (1951) 293–314.
85. H. MROSE, *Angew. Meteorol.*, 2 (1955) 147–151.
86. H. WÖRNER, *Z. Meteorol.*, 4 (1950) 103–110.
87. D. T. MARING, *Monthly Weather Rev.*, 25 (1897) 485–490.
88. C. F. MARVIN, *U.S. Weather Bureau, Circ.* (1941).
89. C. F. BROOKS AND E. S. BROOKS, *J. Meteorol.*, 4 (1947) 105–115.
90. B. FOSTER AND L. W. FOSKETT, *Bull. Am. Meteorol. Soc.*, 34 (1953) 212–215.
91. A. ÅNGSTRÖM, *Quart. J. Roy. Meteorol. Soc.*, 50 (1924) 121.
92. S. FRITZ AND T. H. MACDONALD, *Heating and Ventilating*, 46 (1949) 61–64.
93. J. GEORGI, *Strahlentherapie*, 31 (1929) 368–378.
94. H. BERG, *Geofis. Pura Appl.*, 13 (1948) 3–12.
95. J. C. THAMS AND E. ZENONE, *Landwirtsch. Jahrb. Schweiz*, 66 (1952) 139–180.
96. W. MÖRIKOFER AND E. NAGEL, *Verhandl. Schweiz. Naturforsch. Ges.*, 132 (1952) 112–113.
97. J. N. BLACK, C. W. BONYTHON AND A. PRESCOTT, *Quart. J. Roy. Meteorol. Soc.*, 80 (1954) 231–235.
98. R. W. HAMON, L. L. WEISS AND W. T. WILSON, *Monthly Weather Rev.*, 82 (1954) 141–146.
99. H. HINZPETER, Die Berechnung der Globalstrahlung aus der Sonnenscheindauer, *Obs. Potsdam, Internal Rept.*

100. E. NAGEL, *Ann. Schweiz. Ges. Balneol. Klimatol.*, (1955/56) 27–35.
101. J. C. THAMS, *Arch. Meteorol. Geophys. Bioklimatol. Ser. B*, 6 (1955) 417–430.
102. E. NAGEL, *Ann. Schweiz. Meteorol. Zentralanstalt*, 92, No. 8 (1955) 8/2–8/26.
103. A. ÅNGSTRÖM, *Arkiv Geofysik*, 2 (1956) 471–479.
104. J. GLOVER AND J. S. G. MCCULLOCH, *Quart. J. Roy. Meteorol. Soc.*, 84 (1958) 172–175.
105. F. STEINHAUSER, *Wetter und Leben*, 8 (1956) 1–12.
106. N. ROBINSON AND M. SEMEL, *Bull. Res. Council Israel, Sect. F*, 7 (1958) 193.
107. *Farb- und Filterglas für Wissenschaft und Technik*, Jenaer Glaswerk Schott & Gen., Mainz, No. 3651 (1954).
108. *Glass Color Filters*, Corning Glass Works, Corning, N.Y. (1959).
109. *Kodak Wratten Filters*, Kodak Publ. No. B-3, Eastman–Kodak, Rochester, N.Y. (1957).
110. K. H. BACHMANN, *Z. Meteorol.*, 4 (1950) 176–179.
111. *Schott-Interferenzfilter*, Jenaer Glaswerk Schott & Gen., Mainz.
112. *Balzers Interferenzfilter*, Balzers A.G., Balzers, Liechtenstein.
113. I. DIRMHIRN AND F. SAUBERER, *Wetter und Leben*, 10 (1958) 159–163.
114. A. ÅNGSTRÖM AND A. J. DRUMMOND, On the influence of thickness and temperature on the 'cut-off' characteristics of glass filters, Symposium on Radiation, Oxford 1959. *UGGI Monograph* No. 4, p. 9 (1960).
115. K. YA. KONDRATYEV, *Radiant Solar Energy* (in Russian), Leningrad (1954), p. 143.
116. IGY Instruction Manual, Part VI: Radiation Instruments and Measurements, *Annals of IGY*, V (1957), Pergamon, London, p. 401.
117. P. BENER, *Phys. Meteorol. Obs. Davos, Internal Rept.*, No. 198 (1956).
118. P. BENER, *Phys. Meteorol. Obs. Davos, Internal Rept.*, No. 199 (1957).
119. P. BENER, *Sci. Proc. Intern. Assoc. Meteorol.* (Xth Assembly IUGG, Rome, 1954) (edited by R. C. Sutcliffe), Butterworths, London (1956), pp. 543–548.
120. P. BENER, *Phys. Meteorol. Obs. Davos, Internal Rept.*, No. 199 (1957).
121. R. STAIR, *J. Res. Natl. Bur. Std.*, 40 (1948) 9–19.
122. R. STAIR, *J. Opt. Soc. Am.*, 43 (1953) 971–974.
123. N. RODSKJER, *Arkiv Geofysik*, 2 (1955) 377–384.
124. E. HILTROP AND K. H. GROSNICK, *Z. Meteorol.*, 9 (1955) 366–368.
125. W. ESCHKE, *Angew. Meteorol.*, 2 (1956) 237–240.
126. R. SCHULZE, Laufende Registrierung der UV-A- und UV-B-Strahlung, *Intern Radiation Conference Rome, 1954* (mimeographed, not printed).
127. R. SCHULZE, *Ann. Meteorol.*, 4 (1951) 176.
128. R. FLEISCHER, *Geofis. Pura Appl.*, 26 (1953) 172.
129. H. WÖRNER, *Z. Meteorol.*, 9 (1955) 248–250.
130. *Z. Elektrotech.*, 7, No. 13 (27 March) (1954) 36.
131. G. A. VESZI, *J. Brit. Inst. Radio Engrs.*, 13 (1953) 183–189.
132. E. S. TRICKETT AND L. J. MOULSLEY, *J. Agr. Eng. Res.*, 1 (1956) 1–11.
133. *Weston Light Measuring Instruments*, Weston Company, Circ. B–22–G (1958) and private communication (1959).
134. K. YA. KONDRATYEV, *Radiant Solar Energy* (in Russian), Leningrad (1954), pp. 64–65.
135. P. COURVOISIER, *Verhandl. Schweiz. Naturforsch. Ges.*, 130 (1950) 152.
136. W. MÖRIKOFER, The determination of the radiation balance of the earth, *9th General Assembly IUGG, Brussels 1951*, Meteorol. Assoc., p. 207.
137. P. COURVOISIER, *Phys. Meteorol. Obs. Davos, Internal Rept.*, No. 190 (1956).
138. J. T. GIER AND R. V. DUNKLE, *Trans. Am. Inst. Elec. Engrs.*, 70 (1951) 1–7.
139. R. SCHULZE, *Geofis. Pura Appl.*, 24 (1953) 107.
140. R. SCHULZE, *Ann. Meteorol.*, 6 (1953/54) 127.
141. W. SCHÜEPP, *Arch. Meteorol. Geophys. Bioklimatol. Ser. B*, 1 (1949) 257.

142. F. VOLZ, *Arch. Meteorol. Geophys. Bioklimatol. Ser. B*, 10 (1960) 100–131.
143. P. VALKO, *Arch. Meteorol. Geophys. Bioklimatol. Ser. B*, 12 (1963) 458–473.
144. L. B. ALDRICH, *Smithsonian Inst. Misc. Collections*, 111, No. 12 (1949) 1–4.
145. A. J. DRUMMOND, Eppley Laboratories, Newport, R.I., private communication.
146. H. J. DE BOER, Netherlands Royal Meteorological Institute, De Bilt, private communication.
147. F. MÖLLER, *WMO Bull.*, 6, No. 1 (1957).
148. W. MÖRIKOFER, *WMO Bull.*, 6, No. 4 (1957).
149. P. BENER, *Contract AF61(052)–54*, Tech. Summary Rept. No. 1, Dec. 1960.

Chapter 8

THE APPLICATION OF SOLAR ENERGY

In this chapter we shall consider two applications of solar energy, namely, conversion into heat and direct conversion into electrical energy. Because of lack of space, applications such as those to building science, physiology, biology, and so on, will be omitted. For an extensive bibliography on the various applications of solar energy, see [26]. For a discussion of bioclimatological applications, see [24].

1. CONVERSION INTO HEAT

Low temperature devices without optical concentration: flat plate collectors

Devices of this kind may be used to heat a fluid, such as water, air, etc. The heat thus gained can then be used directly in a central heating system, refrigeration plant (with suitable modifications), drying of agricultural products, distillation of water, and so on. The advisability of using solar energy for such purposes depends on economic considerations since such devices necessarily compete with installations using the other available sources of energy. For example, in the case of a hot water supply which employs solar radiation as the source of energy without any auxiliary equipment for heating in the absence of the radiation, the basic investment and capital depreciation of a solar heater compare favourably with gas or electrical heating whenever these sources of energy are used [1]. In the case of a solar heater incorporating an auxiliary electrical heater (which is brought into operation in the absence of solar radiation) there is of course the additional cost of installation to be considered, as well as the cost of electricity. In general, the choice of a solar heater must be based on a statistical study of the amount of solar radiation reaching the particular locality over a long period of time. Such statistical considerations may show that solar heaters are not economically advisable for the particular place. There are however methods which may be used to economize on the exploitation of the auxiliary source of energy; they involve the so-called 'solar switch' [1].

Solar water heaters

A solar water heater consists basically of two parts, namely, a collector which absorbs the solar radiation and a reservoir to store the heated fluid leaving the collector (Fig. 8.1). In the simplest case the collector is fixed and does not make use of optical concentration. The design of the system is such as to use the principle of multiple circulation to the best advantage. Thus, the temperature difference between the fluid in the collector and in the storage tank, and the difference in the height of the collector and the tank, give rise to this circulation.

The type of collector illustrated in Fig. 8.1 is known as the flat plate collector and makes use of both direct and diffuse radiation. The collector takes the form of a suitable arrangement of pipes or ducts leading the liquid from the entrance E_1 at the lower end of the collector to the exit E_2 at the upper end. The tubes or ducts are embedded in a thermally insulating material, with the side facing the sun blackened and covered by one or more flat sheets of glass or plastic. This arrangement gives rise to an effect known

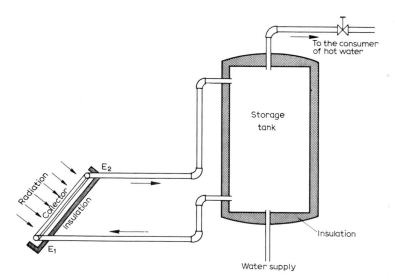

Fig. 8.1. Solar fluid heater (schematic).

as the 'greenhouse effect' which depends on the selective transmissivity of the glass.

Fig. 8.2 shows the transmissivity of ordinary glass as a function of λ. Curve I shows the spectral curve of the solar radiation at the earth's surface. According to Wien's displacement law (cf. Chapter 1), the wavelength corresponding to the peak of the spectral curve should increase with decreasing temperature. Since the absolute temperature of the collector is lower

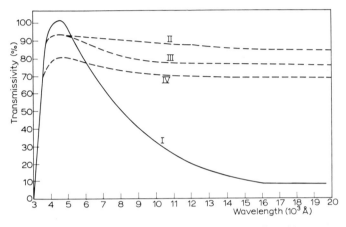

Fig. 8.2. Spectral transparency of regular glass (Israeli made).
Curve I   : Spectral curve of solar radiation.
Curve II : Glass plate 2 mm thick.
Curve III: Glass plate 4 mm thick.
Curve IV: Two glass plates 2 mm thick with 3 mm distance between them.

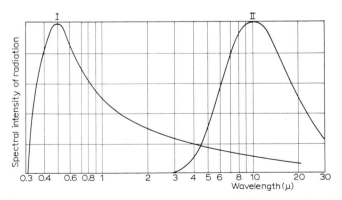

Fig. 8.3. Spectral curves of solar radiation (6000°K) (curve I) and of a collector surface (373°K) (curve II).

References pp. 342–343

than that of the sun by a factor of 16, it follows that λ_m will be larger by the same factor. Fig. 8.3 shows spectral curves for $T=6000°K$ (curve I) and for $T=373°K$ (curve II). Comparison with the transmissivity curve shown in Fig. 8.2 shows that the energy represented by curve II in Fig. 8.3 will be completely absorbed by the glass (this is the so-called dark radiation). The absorbed energy heats the glass and this may be used to achieve higher temperatures (when the collector surface is suitably shielded from wind).

Let us now consider the heat balance of a collector of this type in greater detail.

The heat gained by the collector consists of two parts, namely, direct solar radiation I_y corresponding to the given elevation of the sun $\gamma_\odot$ and the scattered (sky) radiation D_y. The amount of direct solar radiation incident normally on a collector in a time t is given by

$$Q_1 = I_y A t \tag{8.1}$$

where A is the area of the collector. When the angle of incidence varies from 0 to i this relation must be modified to read:

$$Q_1' = I_y A t \; \overline{\cos i} \tag{8.2}$$

where $\overline{\cos i}$ is the mean value of cos i during the time t. However, only a

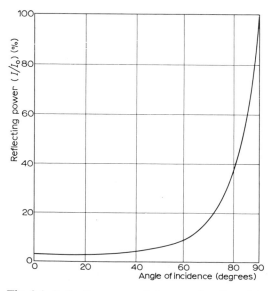

Fig. 8.4. Reflecting power of glass as a function of the angle of incidence [2].

fraction Q_1'' of Q_1' reaches the blackened surface of the collector and is absorbed by it, so that the actual heat absorbed is

$$Q_1'' = \alpha_1 Q_1' \qquad (8.3)$$

The proportionality constant α_1 is a function of the angle i, the number of plates employed, and their thickness and optical quality.

There are three ways in which the absorbed radiation depends on angle i.

Firstly, the reflectivity of the glass is a function of this angle and the reflected energy is, of course, lost and cannot be used for heating. The 'reflecting power' of the glass is given by

$$I_r/I_\gamma = \frac{\sin^2(i-r)}{2\sin^2(i+r)} + \frac{\tan^2(i-r)}{2\tan^2(i+r)} \qquad (8.4)$$

where I_γ and I_r are the intensities of incident and reflected radiations, and r is the angle of refraction which is related to the refractive index n of the glass by $n = \sin i/\sin r$. It is assumed that the incident radiation is unpolarized. Since for normal incidence $i=0$, Eq. (8.4) reduces to

$$I_r/I_\gamma = (n-1)^2/(n+1)^2 \qquad (8.5)$$

For normal glass with $n=1.5$ the reflection losses at normal incidence constitute 4 % of the incident energy.

The dependence of the reflecting power on the angle of incidence is illustrated in Fig. 8.4 from which it is clear that the reflecting power remains practically constant up to about $i=50°$ and then rises rapidly with increasing i. This fact has to be taken into consideration in defining the tilt of the collector. When there are several glass plates covering the collector, multiple reflections between the glass surfaces have to be allowed for.

Secondly, the magnitude of the angle of incidence i has an effect on the absorption, since the optical path is a function of i. Hence, according to Bouguer–Lambert's law

$$I = I_0 \exp(-a \cdot d) \qquad (8.6)$$

where I_0 is the intensity of the incident solar radiation, I is the intensity transmitted through a thickness d of the glass and a is the absorption coefficient. The dependence of the transmittance of i for 1, 2 and 3 glass plates is shown in Fig. 8.5 from which it is clear that the transmittance remains practically constant up to $i=30°$ and then decreases very rapidly with increasing i.

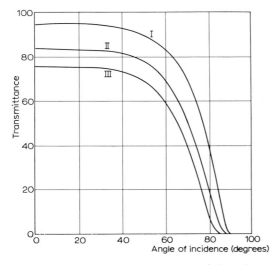

Fig. 8.5. Dependence of transmittance of glass plates on the angle of incidence [2]. (I) 1 plate, (II) 2 plates, (III) 3 plates.

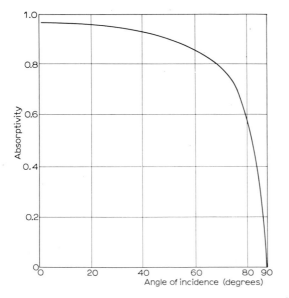

Fig. 8.6. Dependence of absorptivity of a dull black surface on the angle of incidence [2].

Although special kinds of surface-treated glass designed to reduce reflectivity and possessing minimum absorption are available commercially, the above factors must be taken into consideration because solar heaters have to be constructed as simple and cheaply as possible, using locally available materials.

Finally, the absorptivity of a blackened surface depends on the angle of incidence i. This is shown in Fig. 8.6 from which it is clear that the absorptivity remains practically constant up to about $25°$ and then decreases very rapidly beyond $i = 70°$.

So far, we have been concerned with direct solar radiation. The contribution due to sky radiation at the collector is given by

$$Q_2 = D_y A t \cos^2 \frac{\varepsilon}{2} \tag{8.7}$$

where D_y is the intensity of the sky radiation on a horizontal plane and ε is the tilt angle of the irradiated surface, *i.e.* the angle above the horizon [3]. For a uniform sky luminance, it turns out that D_y is only a function of $\gamma_\odot$ and time (for given atmospheric conditions) [4]. It follows that only D_y has to be averaged over the time t, so that the mean value of Q_2 is given by

$$Q_2' = D_y A t \cos^2 \frac{\varepsilon}{2} \tag{8.8}$$

A fraction α_2 of Q_2' is absorbed by the blackened surface so that the final energy reaching the collector is

$$Q_2'' = \alpha_2 Q_2' \tag{8.9}$$

Both Q_1'' and Q_2'' are at our disposal for the purpose of heating.

Let us now consider the heat losses. In addition to the heat lost by reflection at the glass plates and the blackened surface, and by the absorption in the glass itself, there are also the other losses. They are due to dark radiation and conduction and convection, both regular and forced by wind. Heat losses from the insulated sides can be diminished by using an improved insulating material, or by thicker layers of it, and will not be considered here.

To begin with let us discuss the dark radiation from the blackened surface. According to the Stefan–Boltzmann law, the loss by radiation is given by

$$q = \alpha\sigma(T_c^4 - T_s^4) \tag{8.10}$$

where α is the absorption coefficient of the radiating body and is equal to unity for a perfect black body; σ is the Stefan–Boltzmann constant (5.73×10^{-5} erg $\cdot$ deg^{-4}), T_c is the absolute temperature of the radiator (collector) and T_s is the absolute temperature of the surroundings. For a freely exposed body the latter temperature is the temperature of the upper layers of the atmosphere (on a cloudless day). For an overcast sky this temperature is equal to the temperature of the clouds.

It is evident that heat losses by dark radiation are rather high. If, for example, the temperature of a freely exposed and unprotected collector surface is 373°K, the amount of energy emitted is 0.11 watt $\cdot$ cm^{-2}, assuming the emitter surface to be a perfect black body. Since the insolation is of the order of 0.13 watt $\cdot$ cm^{-2} at normal incidence, it is clear that 100°C (373°K) is a difficult goal to achieve for a freely exposed and unprotected black body exposed to solar radiation, even when all other losses are ignored.

In order to increase the absorptivity of the irradiated body throughout the spectrum, the body is painted dull black. This however increases the emissivity of the body, which according to Kirchhoff's laws (see Chapter 1) is given by

$$E_{em}/E_{ab} = f(\lambda, T) = E_B \tag{8.11}$$

where E_{em} is the emissivity and E_{ab} the absorptivity of the body, and E_B is the emissivity of a perfect black body. It is evident that an increase in the absorptivity and the emissivity in the same proportion for all wavelengths will not cause a rise in the temperature of the body.

As shown by Fig. 8.3, there is a considerable difference in the spectral composition of the incoming solar radiation and that of the dark radiation. If the upper surface of the collector is a good absorber of solar radiation but a bad one of the long-wave dark radiation, it will also be a poor emitter of dark radiation. This idea has been utilized by Tabor to prepare a special coating [5,6].

In computing the losses by dark radiation it must be remembered that the glass plate is located at a certain distance from the blackened plate. For this type of arrangement we have approximately

$$q_1 = \frac{T_c^4 - T_{G_1}^4}{E_c^{-1} + E_{G_1}^{-1} - 1} \tag{8.12}$$

where q_1 is the heat lost by dark radiation from the collector in the direction of the neighbouring glass plate (G_1), T_c is the absolute temperature of the collector, T_{G_1} is the absolute temperature of the glass plate, E_c is the emissivity of the collector surface and E_{G_1} is the emissivity of the glass plate. With decreasing E_c both E_c^{-1} and q_2 will clearly increase. This also applies to E_{G_1}. The first glass plate G_1 will totally absorb the long-wave radiation emitted by the collector surface. This will raise the temperature of the glass plate and turn it into a dark radiator.

Usually, a second or even a third glass plate is added. Thus, if another plate G_2 is present, then

$$q_2 = \frac{T_{G_1}^4 - T_{G_2}^4}{E_{G_1}^{-1} + E_{G_2}^{-1} - 1} \tag{8.13}$$

For the uppermost glass plate

$$q_n = E_{G_n} \sigma (T_{G_{n-1}}^4 - T_s^4) \tag{8.14}$$

As already mentioned, T_s may be very low so that the temperature of the uppermost glass plate will greatly decrease, and this will affect the temperature in the direction of the collector surface.

All losses given by Eqs. (8.12)–(8.14) are pure radiation losses. Eqs. (8.10)–(8.14) are valid for surfaces which may be regarded as black for the radiation under consideration. However, in the case of polished or oxidized metal surfaces other losses are found to hold. For further details, see [25].

In addition to the losses mentioned above, heat is lost by convection in the space between the collector and the adjacent glass plate and in the space between the glass plates. Losses are also caused by absorption in the glass plates. In estimating the losses by convection in the air spaces we must consider the tilt of the collector. The losses in the space between the collector and the first glass plate are given by

$$q_3 = \alpha_3 (T_c - T_{G_1}) \tag{8.15}$$

where α_3 is given by

$$\alpha_3 = K(T_c - T_{G_1})^{\frac{1}{4}} \tag{8.16}$$

and K is a function of the tilt. The exact form of the latter function is still not known with sufficient accuracy. It appears however, that K decreases with increasing angle between the collector and the horizon.

For two neighbouring glass plates G_1 and G_2:

$$q_4 = \alpha_4 (T_{G_2} - T_{G_1}) \tag{8.17}$$

and so on.

The combined heat losses in the outward direction give rise to a temperature gradient, and the steeper this gradient the greater are the losses and the smaller is the efficiency of the collector.

Consider now the uppermost plate whose surface is the only one in direct contact with the surrounding air, and is therefore exposed to wind. For a body at a temperature T_B, which is freely exposed to air at a temperature T_A, the convection losses are given by

$$q_5 = \alpha_5 \, (T_B - T_A) \tag{8.18}$$

where α_5 is a function of the wind velocity. In order to calculate the effect of the wind velocity on the cooling of a flat body, it will be convenient to assume that the wind direction is parallel to the surface of the body. It is then necessary to determine the maximum velocity at which the air flow can still be regarded as laminar. Fig. 8.7 shows that laminar flow may be assumed for Reynolds' numbers $R_e \leqslant 2.80 \times 10^6$.

The kinematic viscosity v of air at room temperature is given by

$$v = \mu/\varrho = 0.145 \text{ cm}^2 \cdot \text{sec}^{-1} \tag{8.19}$$

where μ is the coefficient of viscosity of air and ϱ is its density. It can be shown that

$$R_e = \varrho v l / \mu \tag{8.20}$$

where v is the air velocity (in cm/sec) and l is the length of the flat body

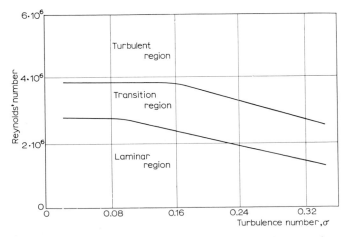

Fig. 8.7. The Reynolds' number as a function of turbulence number.

(in cm). When $R_e = 2.80 \times 10^6$, it is found that $lv/0.145 = 2.80 \times 10^6$ so that $l = 4.06$ m for $v = 10$ m/sec. It follows that the air flow may be looked upon as laminar for wind velocities of up to 10 m/sec and parallel to a flat body having a length of 4 m (in the direction of the wind). The error introduced by assuming laminar air flow is quite small even for plates longer than 4 m. Under these conditions the heat losses caused by the wind are given by

$$q_6 = N_u \lambda_A (T_B - T_A) \qquad (8.21)$$

where q_6 is the amount of heat transferred per unit time from a strip of a flat body of a unit width and parallel to the air flow, N_u is Nusselt's number ($0.585 R^{\frac{1}{2}}$) and λ_A is the thermal conductivity of the air. q_6 is equal to the heat removed by the wind from the upper surface of the glass. Convection losses at zero wind velocity must be added to this.

It is evident from the above discussion that the overall heat balance of a flat collector is a rather complicated two-way process, occurring both in the inward and outward directions.

The total amount of heat absorbed by the upper surface of the collector, beneath which the fluid to be heated is made to flow, is one of the factors determining the efficiency η_c of the collector. The efficiency can be defined in two ways. The first definition is

$$\eta_c = 100 \, (E_1/E) \qquad (8.22)$$

where E_1 is the amount of heat absorbed by the blackened surface and E is the amount of heat falling on the uppermost plate in the same time interval. We shall refer to this as the 'calorific efficiency'. The 'thermal efficiency' η_c' is on the other hand given by

$$\eta_c' = m\overline{C}T/E_1 \qquad (8.23)$$

where $\overline{C}$ is the mean specific heat of the fluid to be heated, m is the mass of the fluid and T is the temperature rise during one passage through the collector. The latter efficiency is a measure of the exploitation of the heat absorbed by the blackened surface for the actual heating of the fluid. The exploitation of the absorbed heat depends on the following design characteristics of the collector.

(1) The ratio of the area in contact with the flowing fluid which is to be heated and the total irradiated area. More fluid can be heated by increasing this ratio. The spaces of the irradiated area between the tubes or ducts carrying the fluid may be at a higher temperature than that of the tubes or

ducts, thus causing heat flow towards the tubes. Large spaces will cause heat losses by convection, conduction and dark radiation.

(2) The thickness of the fluid layer inside the tubes. There is an optimum thickness, and if this optimum is exceeded the absorbed heat will not penetrate deep enough to heat all the fluid in the tube. If the optimum is not reached, the temperature of the sides of the tubes embedded in the insulating material will rise, with a resulting undesirable heat loss.

(3) The heat conductivity of the tube or duct material. Multiple circulation brings into the tube new fluid from the storage tank to be reheated. This fluid flows with a certain velocity and the higher the conductivity the larger the amount of heat which is transferred to the fluid in each passage through the collector.

(4) The flow velocity of fluid inside the tube. If this velocity is too large, only a part of the absorbed heat is transferred to the fluid and the collector remains too hot, which again gives rise to heat losses. If the velocity is too low then the number of circulations per unit time becomes smaller which gives rise to a reduction in the 'recovery time' of the collector.

(5) The path of the fluid inside the collector. This path should be as straight as possible in order to minimize frictional losses.

The collector is only one part of a solar heater. As mentioned above, it is normally combined with a storage tank and the two are connected by pipes. Other pipes connect the storage tank to the radiators.

The overall efficiency of a solar heater depends on the design of all the parts mentioned above. A detailed design has been worked out by the author in collaboration with Stotter [7].

Concentrators

In order to achieve medium and high temperatures the incoming solar radiation must be concentrated. The application of lenses for these purposes is only of limited interest and will not be considered here. We shall therefore confine our attention to the design and properties of mirrors.

All concentrators suffer from the disadvantage that they only concentrate direct solar energy and the sky radiation is lost. One of the characteristics of the concentrator is its 'concentration rate'. In the ideal case as a perfect mirror, the concentration rate is defined as the ratio of the insolated area normal to the solar beam, to the corresponding area of the image of the sun produced by the mirror. In practice the image consists of a very intense

central area surrounded by a less intense outer region. The practical concentration rate, which is equal to the ratio of the insolated area to the *total*
area of the image is often used in numerical computations. In high temperature calculations only the central area should be used.

Let us now consider two numerical examples of concentration rate.
For a cylindrical mirror having a parabolic cross-section and an aperture of
$2f$, where f is the focal length, the diameter of the sun's image is $f/100$. The
concentration rate is therefore $2f/0.01f = 200$, and the heat concentration is
then $0.133 \times 200 = 26.6$ watt $\cdot$ cm^{-2}, when the solar constant is 0.133
watt $\cdot$ cm^{-2}. On taking into consideration atmospheric losses only, and
assuming that the mirror is perfect, this figure is reduced to 15 watt $\cdot$ cm^{-2}.
It may be shown that this corresponds to a possible temperature of the order
of 1000°C.

For a perfect paraboloidal mirror the concentration rate may be computed from the following rule. If the aperture of the mirror is circular with a
diameter equal to $2f$, then the diameter of the sun's image in the focal plane
is $f/100$ and the concentration rate is $(\pi f^2)/(\pi/4)(f/100)^2 = 40{,}000$.

In practice however the concentration is not uniform across the image
of the sun and the maximum concentration occurs at the centre of the image.
The real image has a diameter at least twice as large as the theoretical diameter and the mean concentration rate is therefore only about 10,000.

Since the concentration rate for the paraboloidal mirror is 200 times
greater than for the cylindrical mirror with parabolic cross-section, the
energy concentrated by the paraboloidal mirror at the centre of the image is
found to be $200 \times 15 = 3000$ watt $\cdot$ cm^{-2} which may be shown to correspond
to a temperature of the order of 4500°C.

Increased concentration causes a decrease in the heat losses and one
can say with some degree of accuracy that the heat losses are inversely
proportional to the concentration rate.

Before dealing with the various types of concentrators there are some
further common factors which must be discussed. To begin with, there is
the loss by the mirrors themselves. Back-coated mirrors suffer from double
absorption loss since radiation traverses the glass twice. Infrared radiation
is affected by this much more than other radiation. There is also absorption
at the reflecting medium itself, especially in the infrared. The cleanliness of
the mirrors is an important problem both from the technological and economic points of view. The absorption is lower for front-coated mirrors, but
such mirrors are expensive, deteriorate with time, and are difficult to clean.

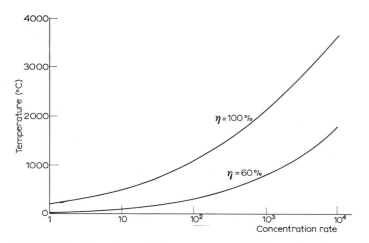

Fig. 8.8. The dependence of the temperature at the focus of a mirror on the concentration rate [8].

For technological and economic reasons the concentrators are usually composed of a large number of small flat mirrors. This type of design does, of course, lead to deterioration in the focusing properties.

In practice, the concentration rate is much lower than the theoretical value and depends on the quality of the mirror, *e.g.* its reflecting power and optical accuracy. If, for example, a parabolic searchlight mirror has a concentration rate of 5000, a parabolic mirror made of reinforced concrete and covered by silvered glass has a rate of only 250–500, which is about the same as for mirrors composed of small plane segments. The approximate relation between the concentration rate and the temperature at the focus of concentrators is shown in Fig. 8.8.

Concentrators for medium temperatures

There are various ways in which concentrators may be used to produce medium temperatures. They are:

(1) A collector with one or more auxiliary concentrators, adjusted to the solar coordinates from day to day.

(2) A concentrator which may be rotated about a fixed horizontal axis together with a collector attached to it.

(3) A concentrator capable of rotation about two axes, one inclined and used for daily rotation and the other for adjustment from day to day.

The collector may be fixed or may rotate with the concentrator.

An equatorial mounting for medium temperatures is too expensive. Method (1) was first used by Shuman [9,10]. The collector was irradiated directly by the sun and the mirrors supplied only an additional amount of reflected energy. A similar arrangement has been used in the USSR by Veynberg [8]. This arrangement is illustrated in Figs. 8.9a and b, in which one mirror faces south and the other mirror faces north. They are both adjusted by rotation about the horizontal axis.

The reflected energy falls on a vertical collector and the mean concentration rate is higher by a factor of about 2.2 as compared with the irradiation by a normal solar beam. It is evident from this arrangement that the concentration rate is a function of time. In order to investigate this device in detail, a working model of it can be constructed and studied throughout the year by means of a special apparatus designed for this purpose [12]. In this way the optimum conditions of operation and the optimum relative dimensions can be determined.

In an arrangement first developed in the USSR [8] in 1935 and used at present in Israel by H. Tabor of the Israel National Physical Laboratory, both the mirror and collector are fixed throughout the day, the latter being arranged above the former.

The first concentrator rotatable about one fixed horizontal axis, which operated for a long time and for which some technical information is available, was built by Shuman in 1911 in Tacony near Philadelphia [9,10].

In order to use the solar radiation intercepted by the concentrator to the best advantage, the concentrator must execute a double motion, namely, a rotation corresponding to the apparent daily path of the sun and a rotation corresponding to the annual change in the coordinates of the sun. This combination of motions may be achieved in various ways. Since the annual coordinates of the sun change slowly, the corresponding axis may be adjusted at short or long intervals, depending on the desired accuracy. The rotation about the daily axis should proceed at a uniform rate.

Owing to the necessary additional degree of freedom, the concentrator must be at a certain height above the ground with the attendant reduction in the rigidity of the construction, as compared with the previous kind, which involved a single fixed horizontal axis, placed close to the ground. Moreover, the elevated concentrator is more sensitive to the effects of wind. These two facts limit the size of this type of concentrator. They have to be built in small units connected together for the daily movement.

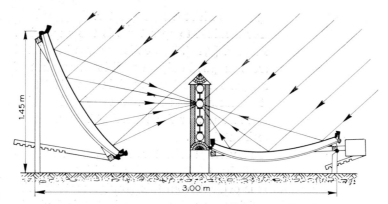

Fig. 8.9a. A system of two adjustable mirrors for the irradiation of a vertical collector [8].

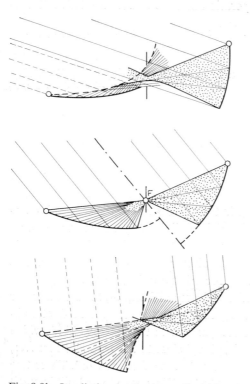

Fig. 8.9b. Irradiation by the two adjustable mirrors in various positions of the mirrors and of the sun [8].

Concentrators for high temperatures: solar furnaces

Two methods may be used to achieve high temperatures. One method is to use a high-quality parabolic mirror on an equatorial mounting, giving the mirror the necessary double motion in order to hold the aperture of the mirror at right angles to the direction of incidence of the solar radiation (Fig. 8.10). Another method is to use a fixed parabolic mirror and to reflect the incident solar radiation by a rotatable flat mirror of suitable dimensions (Fig. 8.11).

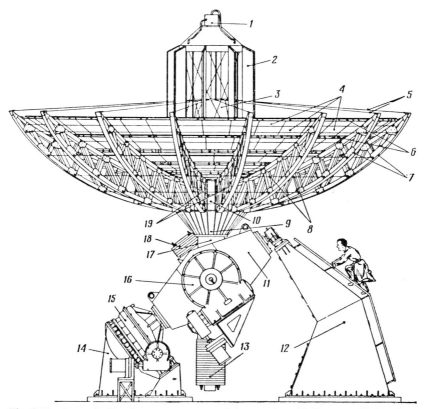

Fig. 8.10. A parabolic mirror on an equatorial mounting. (1) Furnace. (2) Shadings. (3) Conical support. (4) Reflecting mirrors. (5) Tension braces. (6) Mirror frames. (7) Reflector's girders. (8) Fixing braces for girders. (9) Conical support of the central caisson. (10) Central caisson. (11) Earth (world) axis. (12) Northern basis. (13) Contra-balance of the reflector and the frame. (14) Southern basis. (15) Diurnal rotation reduction gear. (16) Inclination reduction gear. (17) Beam. (18) Contra-weight of the inclination reduction gear. (19) Contra torsion braces.

References pp. 342–343

Concentrators possessing high concentration rates are used for special purposes such as the smelting of highly refractory materials, certain chemical reactions, etc. Materials exhibiting special high temperature properties are necessary, such as those used nowadays in modern technical devices, *e.g.* atomic power piles, gas turbines, repulsion engines, special chemical reactors, etc.

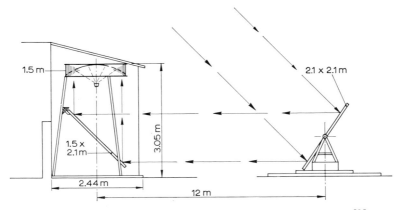

Fig. 8.11. Combined flat and parabolic mirrors for high temperatures [8].

The high-temperature limit of regular furnaces is relatively low (2000°C) owing to the properties of crucible materials, unwanted by-products, oxidation and so on. In electrical furnaces, the upper temperature limit is determined by the evaporation of the heating elements. Although very high temperatures (up to 55,000°C) can be achieved by various methods in the laboratory [11], solar furnaces operating at comparatively low temperatures are more favourable for technological purposes because they are not seriously limited as far as continuity of the operation is concerned. Such solar furnaces can be used under essentially the same conditions during all hours of sunshine. Another advantage of solar furnaces is the fact that they can be operated in a vacuum or 'in the material itself'.

High-temperature solar furnaces are known to have been used for smelting purposes as early as in the 17th century in France, Denmark, Persia and other countries [13]. The temperature achieved by a solar furnace, which cannot be higher than the surface temperature of the sun, is determined by the ratio of the insolated area to the area of the smallest possible cross-section of the solar image. In addition, the temperature depends on the shape and the absorption properties of the target which is to be heated. The mirror need not be as perfect as for astronomical purposes: a searchlight mirror is good enough. A parabolic glass searchlight mirror of good quality has, as mentioned above, a concentration rate of the order of 5000.

A very important factor which has to be taken into consideration is that the sun is not a point source, but possesses finite linear dimensions. This causes image distortions which increase with distance of the solar rays from the centre. It is therefore impossible to get an extended maximum-concentration area in a plane perpendicular to the axis. The distortion depends on the ratio of the focal length to the aperture of the mirror, and both the concentration and the distortion increase with this ratio. It is, therefore, best to use a mirror of medium concentration for the first stage of concentration and another mirror for the second stage.

2. DIRECT CONVERSION INTO ELECTRICAL ENERGY

A high-efficiency solar energy convertor is the silicon photovoltaic cell. It is a semiconductor p–n junction device and the form of it with which we shall be mainly concerned here is the boron-diffused p–n junction developed by Chapin et al. [14,15] which is usually referred to as the solar cell.

Before going into a discussion of the solar cell itself, a short theoretical description of the semiconductor p–n junction and of the p–n junction as a power generator will be given; see also [27,28].

Semiconductors

Semiconductors possess a conductivity which is intermediate between that of conductors and insulators. Semiconductors like silicon or germanium, when extremely pure, exhibit conductivity which is much lower than that of copper, but at the same time considerably higher as compared with insulators such as quartz. In this context the term 'extremely' pure implies that the concentration of impurities affecting the electrical properties of the semiconductor is not greater than one part in 10^9. Such semiconductors are referred to as intrinsic. For practically all semiconductor materials there exist impurities which, when present in concentrations ranging from one part in 10^9 to one part in 10^5, will alter the electrical properties of the semiconductor material as compared with the intrinsic state. The so-called extrinsic semiconductors produced in this way are still extremely pure from the chemical point of view.

The manner in which impurities affect the electrical behaviour of semiconductors may be understood from the following considerations. Semiconductor materials such as silicon are usually employed in their crystalline form. The material must in fact be a single crystal, *i.e.* the atoms are arranged in an orderly crystallographic structure. An individual silicon atom possesses four valence electrons which in the crystalline form of the element are shared with four neighbouring atoms, thus forming a stable crystal lattice (Fig. 8.12).

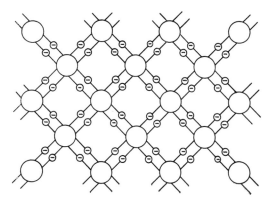

Fig. 8.12. The crystal lattice of silicon, in which each atom shares its four valence electrons with four neighbours.

An impurity atom which replaces a silicon atom in the crystalline lattice may possess *e.g.* five valence electrons (arsenic, antimony or phosphorus). In this case the impurity is left with one extra unattached electron. Such an electron is loosely bound to the impurity atom and the thermal energy is sufficient to remove the electron. The electrons released in this way are relatively free to move and contribute to the conductivity of the crystal in a way analogous to free electrons in metals. A semiconductor in which the impurity atoms donate single electrons, *i.e.* negative charge carriers in order to improve the conductivity, are called n-type semiconductors.

The impurity atom replacing a silicon atom in the lattice may possess only three valence electrons (*e.g.* boron, indium or gallium). In such cases the impurity atom has an insufficient number of electrons to complete the bond with all four neighbouring atoms and may, therefore, capture an electron from a nearby atom to complete the fourth bond. The energy required for this to happen is quite small. However, the captured electron will have left behind another uncompleted bond, which may eventually become completed by another electron leaving behind another uncompleted bond. The position of the missing valence electron is called a hole. In the foregoing picture the uncompleted bond appears and then disappears again and the hole can, therefore, be regarded as being in motion. The hole may be shown to possess an effective positive mass and a positive charge. Semiconductors in which the impurities capture electrons from the surrounding atoms are known as p-type semiconductors.

p–n Junctions

When an n-type semiconductor is placed in contact with a p-type semiconductor the so-called p–n junction is produced. This is usually done within a single crystal sample: on one side of the junction the material contains impurities which make it n-type, while on the other side it contains impurities which make it p-type. There are, of course, considerable concentration differences for the holes and the electrons on the two sides of the junction. These concentration differences give rise to the diffusion of holes into the n-region and of electrons into the p-region. Each flow of carriers leaves behind it a space charge consisting of the corresponding impurity atoms. These space charges produce a built-in electric field. The diffusion effect is eventually balanced by this electric field and an equilibrium state is attained.

Consider now what happens when an external voltage is applied across the junction. Suppose that the positive potential is applied to the p-region and the negative potential to the n-region. The junction is then said to be under a forward bias. The field due to this potential difference is in the opposite direction to the built-in field and the above equilibrium conditions will therefore be altered. This will result in a stronger diffusion of electrons into the p-region and of holes into the n-region, thereby contributing to the forward current. In other words, the positive potential will cause electrons to cross the junction into the p-region and the negative potential will cause holes to diffuse out into the n-region thus giving rise to a forward current. When the polarity of the applied voltage is reversed, the field due to this reverse voltage will be in the same direction as the built-in field, and the result of this will be to cut off the diffusion of carriers from one side of the junction to the other, so practically no current will flow. In other words, the negative potential of the p-region will have no holes to attract from the n-region, and similarly, the positive potential of the n-region will have no electrons to attract from the p-region.

The p–n junction as a power generator

The p–n junction is an essential part of the solar cell. When it is irradiated with solar energy, electron–hole pairs are photo-excited at the front surface of the cell, and when these electron–hole pairs are brought to the junction, the latter acts as a power generator. In order to understand this process let us consider the junction mechanism in greater detail [16].

As was stated above, the equilibrium state in the junction is reached when the electric field created by the diffusion of carriers from one side of the junction to the other is such that it will balance any further carrier diffusion. In the solar cell this is the equilibrium state when the cell is not photo-excited (dark condition). When electron–hole pairs are brought to the junction (as in the case of a solar cell where such pairs are photo-generated), they will be separated by the junction field, upsetting the equilibrium until a new equilibrium is attained. In open-circuit conditions the new equilibrium will give rise to a potential difference across the cell (0.5–0.6 volt in a silicon solar cell). When the cell terminals are shorted a current will flow. When the cell is photo-excited while its terminals are connected to an external load, the new equilibrium will be established after a certain space charge build-up, owing to the separation of the photo-excited pairs. Eventually, a voltage will appear across the terminals, giving rise to a current through the load circuit. It can then be said that the continuous flow of newly arriving photo-excited electron-hole pairs is balanced by 1) additional diffusion current through the junction due to the space charge build-up, and 2) the external load circuit current. Under these conditions the p–n junction will operate as a power generator.

The solar cell

The most commonly used solar cell is the silicon photovoltaic cell which can be made either by diffusion of p-type impurities into a monocrystalline n-type silicon wafer, or by diffusion of n-type impurities into a p-type silicon wafer. Let us consider the first of these.

In practice, the device consists of a diffused junction (Fig. 8.13) carrying back and front contacts. Typical dimensions of such cells are 2×1 cm (length $\times$ width) with a 1 mm front contact along the longer edge. The junction depth and the series resistance of the contact are important in the manufacture of the cell.

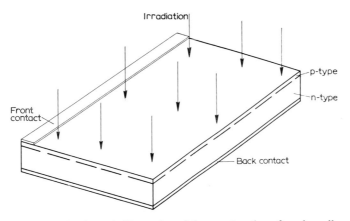

Fig. 8.13. A schematic illustration of the construction of a solar cell.

It is clear that it is most important to get as large a number of photo-excited electron–hole pairs as possible to reach the junction. In order to make this possible, the junction must be very close to the surface of the cell and most of the surface must be free for irradiation. The p–n junction of the solar cell is, therefore, of the shallow-diffusion type, *i.e.* the diffused p-type impurities are allowed to penetrate only a short distance into the n-type silicon material. The final junction produced in this way is parallel to the front photo-active surface very close to it (at a depth of 1–2 microns). This ensures that a sufficient number of photo-excited electron–hole pairs reach the junction [16].

The back contact is made so as to cover the whole back surface, thus reducing the series resistance on this side as much as possible. However, this cannot be done at the front surface where it is very important that the front contact should cover as little as possible of the available area. The optimum width of the front contact strip for a 2×1 cm cell was found to be 1 mm [17]. However, this does not give good enough electrical collection at the photo-active surface since the layer itself acts as a collector when contact is made only at the edge. Since this layer must be made as thin as possible to ensure good penetration yield, it cannot be a very good collecting agent. Additional

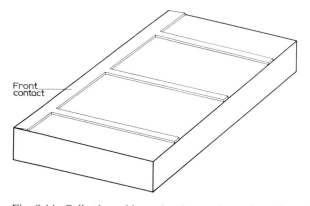

Front
contact

Fig. 8.14. Collecting grids on the photoactive surface of a solar cell.

collecting grids are therefore placed on the front surface (Fig. 8.14) making
the collection distance shorter and thereby improving the solar cell efficiency.

Efficiency of conversion

Silicon is used for solar cells because it is at present the most suitable
semiconductor material. Calculations indicate a maximum theoretical efficiency of about 22 % for the silicon solar cell [17,18]. The efficiency is defined
as the ratio of the electrical power output to the incident radiation, where
the level of the incident radiation is usually taken as 100 mW · cm^{-2}. The
efficiency is very dependent on the utilization of the photon energy in the
generation of electron–hole pairs. For every semiconducting material there
is a minimum energy which is needed for pair generation. This energy is
equal to the energy gap of the material. It can be shown [19] that for every
value of the energy gap there will be a cut-off beyond which the photon
energy will be insufficient to create electron–hole pairs. The smaller the
energy gap the larger the portion of the spectrum which can be utilized.
However, since at certain wavelengths the photons have more energy than is
necessary for power generation, there are energy losses which are larger at
smaller energy gaps. It may be shown [19] that the maximum energy utilization occurs at an energy gap of 0.9 eV (Fig. 8.15). The energy gap for silicon
is 1.07 eV.

Losses due to reflection may be shown to be small [20] and become
appreciable (∼6 %) in the long wavelength range. This low reflection is due

to the high impurity concentration at the photo-active surface which is obtained under proper diffusion conditions.

As might have been expected, the power output is a function of temperature and it may be shown [20,21] that the maximum power output decreases with increasing temperature at approximately 0.02 mW · deg⁻¹.

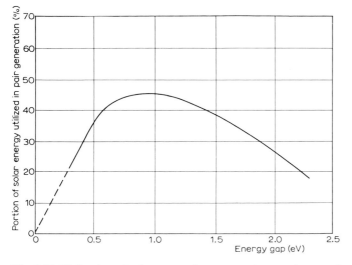

Fig. 8.15. Utilization of solar energy by a p–n pair generator as a function of the energy gap.

The efficiency of a solar cell of the form shown in Fig. 8.13 is about 10%. The average open-circuit voltage is between 0.5 and 0.6 volt, the short-circuit current is between 48 and 55 mA, and the peak efficiency (optimum power output) occurs between 0.37 and 0.41 volt. This is illustrated in Fig. 8.16; curve A shows the current–voltage characteristic for a solar cell of 10.15% peak efficiency and power output of 10 mW · cm⁻² (about 18 mW for a conventional 2 × 1 cm cell). With optimum junction depth and special collector design (*e.g.* the grids of Fig. 8.14) higher efficiency can be achieved. Curve B of Fig. 8.16 shows the current–voltage characteristic of such cells for which the peak efficiency has been increased to 12.1% by the addition of grids etc. Efficiencies of about 13, or even 15%, have been reported [22,23].

A considerable amount of work is being done in order to improve the performance of photovoltaic solar energy converters. In the case of silicon, this means largely technological improvements. Materials other than silicon

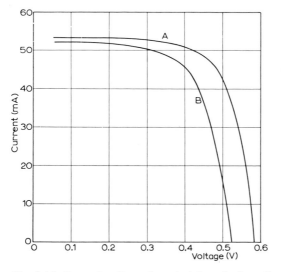

Fig. 8.16. Current–voltage characteristics of solar cells.

are also being investigated, the most promising being antimony. Mention should also be made of other types of cell such as the multilayer cell and the multitransition solar cell [19].

REFERENCES

1. N. ROBINSON AND E. NEEMAN, The Solar Switch, *UN Conference on New Sources of Energy, Rome, 1961*, Preprint E/Conf. 35/S/31.
2. H. C. HOTTEL AND B. B. WOERZ, *Trans. Am. Soc. Mech. Engrs.*, 64 (1942) 91.
3. N. ROBINSON AND D. RECHNITZER, *Bull. Coopération Méditerranéenne pour l'Energie Solaire, Marseille ('Comples Bull.')*, No. 4 (May 1963), pp. 18–24.
4. N. ROBINSON AND L. STOCH, *J. Appl. Meteorol.*, 3 (1964) 179–181.
5. H. TABOR, *Bull. Res. Council Israel, Sect. C*, 5 (1955) 5–27.
6. H. TABOR, *Bull. Res. Council Israel, Sect. A*, 5 (1955) 119–134.
7. N. ROBINSON AND A. STOTTER, *Solar Energy*, 3 (1959) 30–33.
8. V. B. VEYNBERG, *Optics for the Utilization of Solar Energy* (in Russian), Moscow (1959).
9. N. ROBINSON, Solar Machines, in *World Symp. Appl. Solar Energy, Phoenix, Ariz., 1955*, published by Stanford Res. Inst., Menlo Park, Calif. (1956).
10. N. ROBINSON, A Report on the Design of Solar Energy Machines, *UNESCO/NS/AZ/ 141*, Paris (1953).
11. N. ROBINSON, *Tech. Mod.*, 49, No. 4 (1957) 121–132.
12. N. ROBINSON, *Schweiz. Bauztg.*, 71 (1953) 3–9.
13. N. ROBINSON, *Collection Trav. Acad. Intern. Hist. Sci. Paris*, 8 (1953) 510–515.

14. D. M. CHAPIN, C. S. FULLER AND G. L. PEARSON, *J. Appl. Phys.*, 25 (1954) 676.
15. D. M. CHAPIN, C. S. FULLER AND G. L. PEARSON, *Bell Lab. Record*, 33 (1955) 241.
16. M. Y. BEN-SIRA AND B. PRATT, *Semicond. Prod.*, 5, No. 2 (1962) 45.
17. M. B. PRINCE, *J. Appl. Phys.*, 26 (1955) 534.
18. J. J. LOFERSKI, *J. Appl. Phys.*, 27 (1956) 777.
19. M. WOLF, *Proc. Inst. Radio Engrs.*, 48 (1960) 1246–1263.
20. M. B. PRINCE AND M. WOLF, *J. Brit. Inst. Radio Engrs.*, 18 (1958) 583.
21. P. RAPPAPORT, *RCA Rev.*, 20 (1959) 373.
22. M. B. PRINCE, Latest developments in the field of photovoltaic conversion of solar energy, *UN Conference on New Sources of Energy, Rome, 1961*, Preprint E/Conf. 35/S/65.
23. P. A. ILES, *Inst. Radio Engrs., Trans. Military Electron.*, 6 (1962) 5–13.
24. N. ROBINSON, Global solar and sky radiation and their main spectral regions, in *Medical Biometeorology* (edited by S. W. Tromp), Elsevier, Amsterdam (1963), pp. 55–71.
25. N. ROBINSON, *Radiation Properties of Materials*, published by the Engineering Dept. of the Univ. of California, Los Angeles, Calif. (course 150 B, 1961).
26. *Applied Solar Energy Research, A Directory of World Activity and Bibliography of Significant Literature*, 2nd Edition, published by The Association for Applied Solar Energy, Phoenix, Ariz. (1959).
27. W. SHOCKLEY, *Electrons and Holes in Semiconductors*, Van Nostrand, Princeton, N.J. (1950).
28. W. DUNLAP CRAWFORD JR., *An Introduction to Semiconductors*, Wiley, New York (1957).

INDEX